IAEA-TECDOC-1545

Characterization and Testing of Materials for Nuclear Reactors

Proceedings of a technical meeting held in Vienna, May 29–June 2, 2006

IAEA
International Atomic Energy Agency

March 2007

The originating Section of this publication in the IAEA was:

Physics Section
International Atomic Energy Agency
Wagramer Strasse 5
P.O. Box 100
A-1400 Vienna, Austria

CHARACTERIZATION AND TESTING OF MATERIALS FOR NUCLEAR REACTORS
IAEA, VIENNA, 2007
IAEA-TECDOC-1545
ISBN 92–0–103007–X
ISSN 1011–4289

Printed by the IAEA in Austria
March 2007

FOREWORD

Nuclear techniques in general and neutrons based methods in particular have played and will continue to play an important role in research in materials science and technology. Today the world is looking at nuclear fission and nuclear fusion as the main sources of energy supply for the future. Research reactors have played a key role in the development of nuclear technology. A materials development programme will thus play a major role in the design and development of new nuclear power plants, for the extension of the life of operating reactors as well as for fusion reactors. Against this background, the IAEA had organized a Technical Meeting on Development, Characterization and Testing of Materials — With Special Reference to the Energy Sector under the activity on specific applications of research reactors. The meeting was held in Vienna, May 29–June 2, 2006. There was also participation by experts in techniques, complementary to neutrons.

The participants for the technical meeting were experts in the utilization of nuclear techniques namely the high flux and medium flux research reactors, fusion research and positron annihilation. They presented the design, development and utilization of the facilities at their respective centres for materials characterization with main focus on materials for nuclear energy, both fission and fusion. In core irradiation of materials, development of instrument for residual stress measurement in large and / or irradiated specimen, neutron radiography for inspection of irradiated fuel, work on oxide dispersion strengthened (ODS) steels and SiC composites, relevant to future power systems were cited as application of nuclear techniques in fission reactors. The use of neutron scattering for helium bubbles in steel, application of positron annihilation to study helium bubbles in Cu, Ti-stabilized stainless steel and void-swelling studies etc. show that these techniques have an important role in the development of materials for energy sectors. In addition there were brainstorming sessions on the current facilities available and future needs.

This publication presents the summary of the deliberations of the technical meeting which is followed by the presentations by the participants. The IAEA thanks all the experts for their contribution to the technical meeting through presentation of their work, detailed discussions on various aspects of the topic and the manuscript to this publication.

The IAEA officer responsible for this publication was S.K. Paranjpe from the Division of Physical and Chemical Sciences.

EDITORIAL NOTE

The papers in these proceedings are reproduced as submitted by the authors and have not undergone rigorous editorial review by the IAEA.

The views expressed do not necessarily reflect those of the IAEA, the governments of the nominating Member States or the nominating organizations.

The use of particular designations of countries or territories does not imply any judgement by the publisher, the IAEA, as to the legal status of such countries or territories, of their authorities and institutions or of the delimitation of their boundaries.

The mention of names of specific companies or products (whether or not indicated as registered) does not imply any intention to infringe proprietary rights, nor should it be construed as an endorsement or recommendation on the part of the IAEA.

The authors are responsible for having obtained the necessary permission for the IAEA to reproduce, translate or use material from sources already protected by copyrights.

CONTENTS

Summary 1

PRESENTATIONS

Neutron-based techniques for application to advanced energy systems
  materials research and development 15
  R P Harrison
Materials characterization using positron annihilation spectroscopy 25
  G Amerandra, R.Rajaraman, C.S. Sunder, B. Raj
Current activities and future facilities in Canada 37
  A. McIvor
Design features and current status of HTR-10GT 45
  Wang Jie, Huang Zhiyong, Yang Xiaoyong, Shi Lei, Yu Suyuan
The development programme of neutron beam facilities at CARR 63
  Liu Tiancai, Ke Guotu
Residual stress analysis by means of neutron diffraction at
  research reactors — facilities and applications at the HFR 69
  C. Ohms, R.C. Wimpory, D. Neov, A.G. Youtsos
The structural materials of tritium-breeding blankets in
  fusion reactors 87
  J.-L. Boutard
Neutron radiography of advanced nuclear fuels 95
  K.N. Chandrasekharan, H.S. Kamath
Utilization of 14 MW TRIGA research reactor integrated in a structure
  for materials and nuclear fuel characterization and development 109
  C. Paunoiu, M. Ciocanescu, M. Pârvan, M. Mincu, O. Uţă, S. Ionescu
Irradiation devices for fission and fusion materials testing
  in the HFR, Petten 119
  B. van der Schaaf, J. van der Laan
Research reactors in Argentina and their neutrons applications 127
  M. Schlamp
Materials research — A challenge for fission, fusion and
  accelerator research 135
  G. Mank

List of Participants 141

SUMMARY

1 INTRODUCTION

Long term energy security will depend on innovative reactor design and fuel cycles. Further developments in reactor design and development are needed in order to achieve higher burnups in thermal reactors, and develop more efficient fast reactors and associated fuel cycle facilities that utilize a far larger percentage of the fissile and fertile ore. The energy challenge facing the international nuclear industry is to develop the next generation of reactors that will use the large fraction of fertile uranium-238 (over 99% of the uranium in ore), instead of simply using the uranium-235 fraction and treating the remainder as waste. If only half of the available U-238 is used in new types of reactor, that will extend the life of current known uranium reserves by a factor of 60. Development of materials for fusion reactors, is another challenge for the energy sector. Many centuries of electricity production without contributing to global warming is the challenge for which new materials are needed.

Engineering components in all these applications rely on advanced materials for their safe and reliable operation. The crystallographic or chemical structure, hardness, anisotropy in thermal and electrical conductivity, response to external mechanical and thermal loads and to magnetic fields and radiation resistance are some of the properties that determine the suitability of a material for use in future reactor designs.

The demands of quality control on each component to be used in the nuclear energy sector and many heavy engineering sectors are very stringent and are critical to their safe and efficient usage. These, in turn, call for the development and fabrication of components with high quality and reliability. This process demands the characterization of the material, and of various fabricated parts, during the process of fabrication. The testing of these parts after use provides information on their behaviour under various conditions of operation and/or in the diagnosis in case of their failure. Various techniques have been developed to measure material properties at the microscopic and macroscopic level. X rays and particle beams are useful in the study of thin samples. Allied techniques such as positron annihilation spectroscopy (PAS), scanning tunnelling microscopy (STM), electron microscopy (SEM, TEM), and NDT tools, each have a role to play in materials characterization ranging from the atomistic level to the macroscopic scale.

Research reactors (RR) play a crucial role in such developments for testing and characterization of new fuels and structural materials. In-core facilities are needed to provide realistic conditions for testing new fuels and materials. Neutron beams play a unique role both in the characterization and development of materials. Neutrons are a bulk probe, penetrating deep into a sample without damaging the structure of the material under examination. Some of the specific applications of neutron beams are:

- Measurement of residual strain formed during manufacturing processes and in service.

- Measurement of particle size and distribution, void distribution, helium bubbles, phase and texture analysis.

- Transmission techniques such as neutron radiography or tomography.

- Study of surfaces, thin films and buried interfaces.

A Technical Meeting (TM) entitled Development, Characterization and Testing of Materials with special reference to Nuclear Energy was organized under the umbrella of specific applications of RRs. The multidisciplinary character of the meeting with involvement of facility developers, users of RRs and stakeholders in energy and other related areas is evident from the title of the TM. Although main focus was on the RRs, to get a broader perspective related complementary and supplementary techniques were included in the discussion.

1.1 The stakeholders for various facilities

The operators and the user groups working at the research reactors develop instruments or facilities for materials characterization. The agencies interested in using them would be:

- Operators of nuclear power plants who may be looking for information to justify plant life extension or help to address major material issues such as embrittlement, swelling and residual stresses.

- Designers seeking novel technology for example INPRO (International Project on Innovative Nuclear Reactors and Fuel Cycles), GEN-IV (Generation IV Nuclear Energy System).

- Scientists from the fusion reactor community looking for suitable materials and characterization and testing for acceptance criteria.

- Materials scientists, metallurgists, chemists involved in the development of new materials.

The meeting was aimed at bringing together the designers/managers from nuclear power, fusion research, and the facility providers/managers of research reactors for brainstorming discussions on the application and enhancement of available facilities and development of new facilities.

1.2 Nature of the meeting

- Presentation of information on the facilities available.

- Presentations by stakeholders on the needs for materials characterization using RRs.

- Presentations by providers of techniques complementary to research reactor utilization were also included.

1.3 Expected output

- Discussions on present status and future needs.

- Possible collaborative activities in future.

- Collaborative experiments — research projects.

- Impact on technological development/sustainability of improvements in the application of available facilities.

    The result is a publication acting as a guideline or reference for the facility developers and users.

2. PROCEEDINGS OF THE MEETING

Under the regular project activity on specific applications of RRs a Technical Meeting on Development, Characterization and Testing of Materials was held May 29–June 2, 2006 in Vienna. The main focus of the meeting was on the materials related to the energy sector. Participants from Argentina, Australia, Canada, China, the European Commission, France, India, Netherlands and Romania attended the meeting.

The participants gave presentations of their activities related to the theme of the meeting. This covered the facilities available, special features of the instruments and need for materials. The presentations by the participants are included at the end of this summary report. Some of the main points brought out by the participants are presented below.

2.1 Summary of the presentations

R. Harrison, Australian Nuclear Science and Technology Organization (ANSTO), Australia gave an overview of neutron-based techniques viz., neutron diffraction, small angle neutron scattering (SANS), neutron radiography and *their applications for materials characterization like residual stress measurements, high speed powder diffraction-in-situ study of combustion synthesis, and helium bubbles in steel – useful for Fusion reactors*, and study of magnetic materials. Some of the unique facilities for carrying out these studies under extreme conditions of temperature, pressure and magnetic fields existing world-wide were presented.

G. Amarendra, Indira Gandhi Centre for Atomic Research (IGCAR), India, presented an overview of the problems in structural materials in thermal and fast reactors and the need for complete characterization in terms of physical and chemical properties. He introduced the *Positron annihilation techniques and their sensitivity and selectivity to vacancy-defects, highlighting their application to helium bubbles in Cu, Ti-stabilized stainless steel and void-swelling studies in D9 steel (15% Cr, 15% Ni)* etc.

A. McIvor, National Research Council (NRC), Canada, presented some of the salient features of the NRU reactor, built in 1957, which has been a work-horse for decades, leading to many pioneering and novel neutron scattering studies in condensed matter research. He highlighted many technical innovations for materials irradiation and characterization using this reactor, *including a facility for neutron diffraction on irradiated samples.* In view of the age of this reactor, there is now an urgent *need for a new research reactor with superior technical features,* which can serve for future needs.

Jie Wang, Tsinghua University, China, presented the highlights of helium gas cooled 10 MW thermal reactor. The salient features of the oxide fuel in the form of spheres, graphite moderator, control rods and turbo generator were presented. *Experience gained due to long term operation of turbo generator and setting up of a rigid rotor test rig for magnetic bearing studies* were illustrated.

Liu Tianci, China Institute of Atomic Energy (CIAE), China, highlighted the important technical details of China Advanced Research Reactor (CARR), which is currently under development. Various neutron-based techniques that are planned to be set up using this facility were detailed.

C. Ohms, Joint Research Centre of European Commission (JRC) Petten, Netherlands, talked about the measurement of residual stress at the 45 MW RR at Petten. He gave some examples of interesting projects like *the diffractometer capable of holding samples of up to 1000 kg., which was shown with a 600 kg specimen that had a thick cladinglayer of weld material built up over one surface.* JRC, Petten also has developed a *shielding that can cover the sample table and detector to allow diffraction experiments on highly radioactive specimens.*

J.-L. Boutard, European Fusion Developmet Agreement (EFDA)-Garching, Germany, spoke on "Structural Materials of the Tritium-breeding blanket of Fusion Reactors". International Thermonuclear Experimental Reactor (ITER) will include demonstration blanket units for limited breeding of tritium. A future DEMO reactor would include full blankets to produce tritium needed for the reaction. *These components experience an extreme environment bombarded with helium nuclei and 14 MeV neutrons.* Research reactors cannot simulate such flux, but *for materials irradiated using other facilities such as IFMIF, research reactors have* *a role to play in the examination of post-irradiated samples.* He gave *examples of use of SANS technique for samples bombarded with He in an accelerator and the results of the study of He bubbles in steels.*

K.N. Chandrasekharan, Bhabha Atomic Research Centre (BARC), India, spoke on "Neutron Radiography on Advanced Nuclear Fuels" covering facilities at two centres for post-irradiation examination, and fabrication of fresh fuel. *KAMINI (30 kW) began operation in 1996, is a small research reactor dedicated to radiography for the Indian fast reactor programme. It has a hot cell directly above the radiography facility so that irradiated fuel can be lowered and images taken.* A number of features of neutron radiography were discussed highlighting its unique powerful features for both qualitative and quantitative examinations. Examples were given of *cracking in insulation pellets in fuel pencils fabricated for irradiation testing, and quantitative evaluation of central voids examination of homogeneity of Pu oxide in a mixed oxide fuel matrix using the CIRUS research reactor.*

C. Paunoiu, Institute for Nuclear Research (INR), Romania, spoke on "The utilization of the 14 MW TRIGA research reactor". In the Romanian facility there are two reactors in a single pool, one steady-state - a *500 kW TRIGA, and a 20 GW pulsed reactor.* The facility has several installations useful for materials characterization: *in core irradiation of fuel, including the ability to fail fuel, gamma radiography and tomography, underwater radiography that can accommodate active samples, CANDU fuel elements irradiation in pulsed reactor, fuel rod puncture test for fission gas pressure and volume, metallographic and tensile testing of irradiated fuels.* He mentioned about the participation of Romania in the international research effort on *delayed hydride cracking in zirconium.* Examples were given of work on pressure tubes, and also fuel cladding.

B. van der Schaaf, Nuclear Research and Consultancy Group (NRG) Petten, Netherlands spoke on "Devices for fission and fusion materials research". He is involved with the *in-core irradiation of fuels and materials* at Petten. The reactor has a thermal flux of 1.6E14 n/cm^2/s. Depending on the experiment, the flux on the sample can be adjusted using Cd and Hf shields. Examples were given of work on oxide dispersion strengthened *(ODS) steels and SiC composites, relevant to future power systems.* An important area of research for the industry is to reduce the *quantity of long-lived waste.* Petten has worked on a project on *Pu inert matrix fuel that would help this issue.* An example was also given on *graphite development work for the VHTR, a good example of today's research reactors helping to develop tomorrow's power reactors.*

There was some discussion of a *future reactor at Petten- the PALLAS reactor-* a JRC NRG TUDelft and Mallinckrodt initiative with 5E14 n/cm^2/s fast flux and 5E13 n/cm^2/s thermal flux and a nominal power around 40 - 80 MW. The horizon for realization is approximately 2015.

M. Schlamp, Comisión Nacional de Energia Atómica (CNEA), Argentina, gave an "Overview of research reactors in Argentina". TA-8 is the newest reactor, with power 10W, operational since 1998 designed *to test the core in CAREM, an inovative small power reactor being developed in Argentina.* Some examples of work in Argentina were given. *Embrittlement of power plant steels under fast flux and H content in Zr alloys with irradiation damage has also been researched.* Argentina is intending to resume work on the new power reactor - Atucha-2. So research support is needed for that project on fuel development. Argentina favors the construction of a new multipurpose reactor. Presently stakeholders support is being sought.

G. Mank, IAEA, spoke on "Materials Research for Fission Fusion and Accelerators" He described how the global population of research reactors had peaked in the 1970s at 400 but is now 276. With this reduced number of facilities there is still much research that needs to be done. With large spallation neutron sources (SNS) coming on-line, new research can be pursued, but new *materials challenges arise also such as the mercury target. International Fusion Materials Irradiation Facility (IFMIF) presents some significant materials challenges with its power and flux.* The European Spallation Source (ESS) Mg target at 5 MW was even more challenging than Japan Proton Accelerator Research Complex J-PARC, Japan or SNS, USA. *Fusion research will need the materials research capabilities of our reactors* also. FIREX in Japan is the "Fast ignition realization experiment". It is a very high temperature experiment driven by laser. ITER is going to be a step closer to break even in fusion power, and the materials in the first wall and tritium breeding blanket have been described earlier. Following ITER, DEMO will be a power generating fusion reactor, but with flux and power more than 10 times ITER. *There is an essential role for research reactors in support of the next generation of fission and fusion reactors;* if member states identify a common research theme there are strong possibilities for an IAEA programme that will have broad impact in the industry.

V. Inozemtsev, IAEA, spoke about development of power reactor fuels. There is a drive towards increased operating temperature and burn-up of fuel. IAEA CRPs are currently active on three topics: *high burn-up fuel, delayed hydride cracking in fuel cladding, and water chemistry/corrosion.* As *fast reactors are developed, there are basic challenges for materials such as swelling or embrittlement in steels.* Neutron diffraction is a good technique for investigating material properties for thermal and fast reactor development.

A. Soares, IAEA, spoke on the IAEA's work with the research reactor community internationally. His group is responsible for: improving the effective utilization of research reactors, modernization / refurbishment of facilities, new research reactor fuel cycles, and for decommissioning and involved in activities related to HEU to LEU conversion and repatriation of HEU from various countries to Russia. Moving from wet to dry fuel storage, and UMo fuels are two other areas of future activity.

3.   DISCUSSIONS AND OUTPUT

3.1.   Facilities/capabilities available at current research reactors

3.1.1   Neutron beam techniques

3.1.1.1 Neutron scattering techniques

The nature of neutrons makes them an extremely versatile tool in materials science. Unlike X rays, they interact rather with atomic nuclei than with electrons. Their magnetic moments allow them to determine the magnetic structure of materials. Other neutron based techniques like radiography and activation analysis utilize the neutron's nuclear properties rather than its wave nature. A number of neutron scattering techniques used in materials research for advanced fission and fusion systems are described below.

(a) Powder diffraction

Analysis of crystalline structures using neutrons is based on the phenomenon of Bragg diffraction. Neutrons impinging on the material under investigation are scattered in preferred directions satisfying the Bragg condition. A diffraction pattern can be recorded using a suitable neutron detector. Relative positions and types of atoms in the lattice can be obtained from the analysis of the recorded spectrum.

Neutrons complement X rays as a materials probe. Neutrons are useful when probing materials containing light elements, and can distinguish neighbouring elements in the periodic table. Neutrons scatter differently from isotopes of the same element. Neutrons are also needed, when activated materials are studied, e.g. *nuclear fuel samples*, because the X ray or γ-radiation emanating from such samples would mask the X ray detector.

Ancillary equipments providing for particular sample environments are available at most neutron diffraction facilities. Such environments would be: high temperature furnaces, low temperature cryogenics, pressure cells, magnetic fields or radiation shielding if necessary. Changing the ambient conditions facilitates the study of phase transitions and magnetic structures. It also allows for testing materials under operational conditions.

Important assets for such powder diffraction facilities are high neutron flux on the specimen and instrumental resolution. The incoming flux is largely determined by the neutron source, but also by the neutron optical equipment installed at the facility (i.e. focusing monochromator). The resolution (at a RR) does rather depend only on the neutron optical equipment of the facility. Another important aspect is the quality/efficiency/resolution of the neutron detector.

Desires: higher flux sources (a large increase may not be realistic for RRs); optimization of neutron optical equipment, tailored to purpose; wide angle good resolution high efficiency detectors; fast detector electronics; improved, combined ancillary equipment; possibility for in-situ, real time experiments; scientific and technical staff for optimized exploitation of facilities; improved collaboration and communication.

(b) Measurement of residual stress by neutron diffraction

Residual stress measurement facilities in the early days (1980 – 1995) were mostly derivatives of powder diffraction facilities. In fact, in many cases dual purpose facilities had been set up. The main difference between a residual stress and a powder diffractometer are the specimen positioning table and the beam defining optics applied in stress diffraction.

Residual stress analysis by neutron diffraction is based on Bragg diffraction, as is powder diffraction. Stresses change the lattice spacing in a crystalline material (strain) and this change is quantified by measuring the shift in the diffraction angle. By measuring strains this way in several directions, stresses can be derived based on the generalized Hooke's law.

Today there are 20-30 residual stress measurement facilities based on neutron diffraction around the world, many of which belong to the second and third generation facilities designed and optimized solely for this particular application.

At the best facilities in operation today, steel specimens of up to 5 cm thickness can be investigated. The best possible resolution in space is less than 1 mm^3.

There are many possible applications in the area of nuclear energy. *Related to the safety of present-day installations, a lot of work is done on welding stresses in steel components.* Many investigations on Zr-alloys have also been performed at various places. For materials and fabrication processes qualification for future installations (Gen-IV, fusion), the applications would be in principle very similar, but they would certainly comprise stress analysis at high temperatures or analyzing the impact of irradiation.

(c) Small angle neutron scattering (SANS)

With small angle neutron scattering, because the neutron beam is scattered at a small angle the recorded patterns can provide information at larger structures, molecules, clusters etc. While traditional diffraction examines structures in the length range of 0.2 nm (i.e. atomic distances) SANS can look at structures of a few hundred nanometres: larger molecular level structures such as polymers, micelles, precipitate formation in metallurgical samples etc.

Unlike the other techniques mentioned here SANS requires the application of cold neutrons, i.e. neutrons with sub-thermal energies. As these are present in the thermal spectra provided by the source itself only in small amounts, SANS facilities are normally equipped with cold neutron sources.

(d) Texture analysis

With a diffraction technique where the sample is rotated around a hemisphere, the data can be compiled into pole figure showing the texture (preferred orientations of the crystal grains) within the sample.

3.1.1.2 Neutron radiography

A transmission technique where neutrons pass through a sample and are captured on film or an electronic detector. Neutrons can pass through dense materials and are sensitive to light atoms so images that are not possible with other techniques can be made.

3.1.1.3 Neutron activation analysis (NAA)

A technique that can measure very small amounts of impurities in materials, in parts per billion. After irradiation, trace elements become radioactive and their characteristic gamma rays can then be detected. This particular technique can be applied at neutron beams, but the activation of specimens can also be performed in-core.

3.1.1.4 Neutron beam based positron beam generation

Since the available flux of positrons from radioactive sources is limited, neutron induced positron emission from certain elements like Cu and Co offers novel possibilities of materials characterization such as positron diffraction, positron re-emission radiography etc. In view of the sensitivity of positrons to vacancy-defects and voids, this offers *a means of characterization of the early stages of helium bubble formation*, which is complementary to SANS.

3.1.2 In-core techniques and applications

3.1.2.1 Fuel irradiation facilities

The following are required as part of facilities associated with fuel irradiation:

- Monitoring behaviour of fuel during irradiation (predominantly temperature).

- Conditioning for post-irradiation examination (PIE).

3.1.2.2 Irradiation facilities for structural materials

Irradiation of structural materials in-core and subsequent post-irradiation examination (PIE) require significant design considerations and facilities. Such testing is not possible at all RRs as it generally requires open type cores with space for loops etc. Similarly, PIE requires extensive facilities and is not something that can be easily (or indeed cheaply) added on to an existing facility without careful consideration and a specific need.

The following list gives examples of the types of equipment required for irradiation facilities:

- High pressure and temperature loops.

- Monitoring and controlling systems (pressure, temperature, flux etc.).

- Facility coolant types (water, gas, sodium, lead, eutectic alloys etc.).

- Stationary and mobile facilities.

- Means of varying the spatial and energy distribution of available neutrons.

Requirements for PIE include:

- Hot cell facilities.

- Wet radiography.

- Neutron beam techniques (shielded facilities generally needed).

- In-pool inspection – optical, ultrasonic.

- Hot cell based materials testing facilities (for post-irradiation examination).

3.2 Requirements for future research and power reactor materials

3.2.1 Role of RRs in non-nuclear field related applications

Neutron-based techniques are being used in a variety of fields such as physics, chemistry, metallurgy, polymer science, biology, medicine etc. The field of magnetism, superconductivity, low-dimensional systems as well as nano-science and technology has been enriched with the availability of neutron scattering techniques, which provide complementary information to other characterization techniques. In many cases, these techniques provide information that cannot be obtained otherwise.

3.2.2 Role of RRs in nuclear technology related applications

Currently, RRs are being extensively used in three broad categories of applications

- Performance evaluation (in-core) — *Fuels, coolants.*

- Irradiation behavior (in-core) — *Structural materials, control rod materials.*

- Characterization (out-of-core) — *irradiated fuels, structural materials.*

These efforts have paved way for the development of newer fuels, structural materials, coolants and neutron absorbers.

The current meeting focused on nuclear technology related applications and the following is the broad summary of the deliberations.

3.2.3 Structural materials issues

Currently operating and planned reactors use Zircaloys, austenitic/ferritic Stainless Steel.

3.2.3.1 Thermal reactors

Currently, thermal reactors use Zircaloy-4 (which has replaced Ziraloy-2) as well as Zr-Nb alloys. Some of the important issues pertain to hydrogen diffusion, hydrides, hydrogen embrittlement, corrosion etc. Unlike other techniques, neutron scattering techniques have very high sensitivity to light impurities like hydrogen in the host matrix and thus are valuable tools to investigate these problems.

3.2.3.2 Fast reactors

Currently, fast reactors with liquid sodium coolant use austenitic steels and developments of these steels (in terms of percentage of Cr and Ni as well as the optimization of minor alloying elements such as P, Si, W, V etc.),.Helium-vacancy interactions, helium bubble formation and consequent void-swelling and dimensional changes are the important materials problems. SANS measurements are valuable in obtaining useful information about the size distribution of precipitates. Residual stress measurements using neutrons can provide valuable information on life extension of the components.

3.2.3.3 Advanced power reactors

It is envisaged that modified austenitic/ferritic steels, oxide dispersion strengthened (ODS) steels or low activation steels will be used in advanced power reactors in order to obtain better performance at higher burnups as well as to reduce long-lived waste material.

With respect to the ODS steels, since oxides in the form of fine precipitates are dispersed in the matrix, SANS can provide quantitative information on the particle size distribution, which is complementary to other established techniques like electron microscopy.

3.2.3.4 Fusion reactors

The extreme conditions of operation of the first wall and tritium breeding blanket in proposed fusion reactors require ingenious design and development of advanced materials. The envisaged materials include graphite, SiC fiber reinforced matrices, modified martensitic steels etc. The acute radiation damage and enhanced helium production in the first wall require extensive studies on point defects, their clustering, helium diffusion, helium-vacancy interactions and helium bubbles to obtain fundamental understanding leading to a better macroscopic prediction of material behavior. Neutron scattering studies can play an important role in this direction, along with other complementary characterization techniques.

3.2.3.5 Fuels

For the next generation of fission reactors the emphasis is on increased temperature and pressure of operation for improved thermodynamic efficiency. New fuel types are also planned for two reasons: to operate safely and effectively under these extreme conditions, and also to enable new fuel cycles that utilize fertile elements such as U-238 and Th-232. For all these advanced fuels, higher burnups are desirable for economic viability. For effective closing of the fuel cycle, the ease of reprocessing is also an important factor.

In addition to power reactors, new research reactors, and upgraded older facilities also require new types of fuel. There is an international effort to develop U-Mo as an alternative fuel to enable reprocessing.

These new designs require a large amount of research into materials if they are to become reality. Research reactors can play a strong role in this research, using both in-core and out-of-core facilities. Testing of new fuels requires in-core facilities that hold fuel samples at conditions of interest. High temperatures and pressures, advanced coolants such as super-critical water, gas or liquid metal and alloys are envisaged for these in-core test facilities.

Following in-core irradiation, fuel characterization can be effectively carried out using neutron beams. Facilities for powder diffraction and residual stress measurement of radioactive materials are available at the Patten and Chalk River neutron scattering laboratories.

3.3   Limitations of current research reactors

While the currently available research reactors have a wide range of capabilities suitable for the examination, characterisation and testing of materials for reactor applications, there are a number of areas where improvements in the physical characteristics or instrumentation would provide significant improvements in the ability to service the future needs of the industry. A selection of areas where improvements are possible are discussed below.

3.3.1 Availability of flux

In general for neutron beam techniques the greater the neutron flux the better, however there are a number of cases where additional flux may affect certain aspects of the measurements being undertaken. For example, in neutron radiography, higher flux results in lower exposure times, but this places significant technical requirements on the shutter mechanism (for the direct technique) for non-radioactive samples. The ability to improve the available flux is dependent on the reactor design, and this may not be feasible in all facilities.

There are a number of methods of providing increased flux. This can be achieved by increasing the reactor power level, improving the efficiency of transfer of neutrons from the core to the instruments for example, using neutron guides with super-mirror surface coatings, thus increasing the signal level at the experimental station and improving the signal-to-noise ratio by lowering the gamma background. Alternatively, the use of improved instrumentation and imaging techniques like real-time imaging in NR will compensate for lower fluxes.

3.3.2 In-core irradiation facilities

In-core irradiation is only possible in reactors which have special provision for sample insertion. Many facilities with compact cores have no space available for flexible experimental plans. Such reactors may have irradiation positions in, for example, a reflector vessel, but these locations will have a reduced fast flux compared with the core. Fast flux (E > 0.1 MeV) is important for materials in many reactor types so such reactors will not be particularly useful in these applications. It is clear that a more flexible core configuration, where experiments and materials can be placed in the high fast flux in-core positions, is a significant advantage for future materials research. There are requirements for testing both non-radioactive and radioactive materials in the in-core positions.

In specifying the core configuration of current RRs it is necessary to allow for the wide range of possible experiments that may be required. Such experiments include materials irradiation (either surveillance coupons or other materials, such as pressure vessel steels), fuel irradiation (for fuel development and qualification) and component irradiation.

3.4 New research reactor proposals

It is clear that new research reactors are essential for the international nuclear industry to progress to the next generation of fission reactors. Many of the challenges for the international fusion research community are related to materials. It is therefore anticipated that neutron scattering will make an important contribution to that field also, as neutrons are a unique probe for materials.

Several countries are constructing new spallation sources for neutron scattering: Japan, USA, UK and China. These countries recognize the importance of advanced materials to a developed economy, and the pivotal role that neutrons can play in the development and assessment of materials and processes. In other countries including China and Australia, new research reactors are under construction in response to very productive older facilities reaching the end of their operation. In those two cases *multipurpose* facilities are being built, recognizing the wide array of benefits that can be derived from such research reactors: medical isotopes, neutron scattering for materials research and neutron irradiation of materials (for example silicon for the electronics industry). However, neither of these reactors has the facility to perform significant (or any) in-core testing due to their compact core designs. These multipurpose facilities are likely to remain relevant, able to adapt to the research needs of their respective nations. There remains however a clear need for research reactors with large flexible cores so that in-core irradiations and testing of fuels and materials can be undertaken.

3.4.1 Innovations in neutron scattering techniques

There will inevitably be developments in neutron based techniques in the future. It is not possible to describe these, however, it is possible to identify areas where enhancements in the current capabilities can be made fairly readily. These include:

- focussing of the neutron beam to increase the flux at the sample,

- the provision of additional shielded sample holders for performing neutron scattering of active materials

4. CONCLUSIONS

There is a declining and ageing population of research reactors. In addition a large amount of research is needed to meet the requirements of materials for advanced nuclear power systems. Based on the discussions during the meeting an increase in the demand for neutron facilities was foreseen.

PRESENTATIONS

Neutron-based techniques for application to advanced energy systems materials research and development

R P. HARRISON

Australian Nuclear Science and Technology Organisation (ANSTO), Australia

Abstract: Research reactors have been utilised in the testing of materials for many years. Neutrons possess a number of advantages over other modalities, for example, in most engineering materials neutrons can travel several centimetres thus enabling information about both surface and sub-surface structures to be obtained. The use of neutrons for characterisation is broadly divided into three areas: (i) neutron scattering, (ii) residual stress measurement and (iii) neutron radiography. Neutron scattering techniques are used to assess microstructure by measuring the scattering of neutrons as they pass through the material (this is a simplification). By examining the scattering patterns it is possible to determine the lattice spacing and crystal structure and also changes in the material as a function of time, during, for example, some thermo-mechanical process. This has been used extensively to study structure/property relationships and is widely reported in the literature. The opportunities exist to perform neutron scattering on samples that are being mechanically or thermally processed in some way. This can provide important information on the development of structure through the fabrication route of the material that is not possible to perform with other methods, for example X rays. Neutrons are useful for examining the stress state within materials. By measuring the lattice spacing (relative to the unstressed state) it is possible to measure residual stresses up to a few centimetres depth in steel and other high density materials. This enables designers to determine the effect of particular processing methods or welding on the stress state within the material; this information being very important in predicting component lifetimes. Additionally, neutrons can be used to determine the microstructural texture of materials. Neutron radiography (NR) has the advantage over conventional radiographic techniques that it is able to detect low density materials inside higher density materials — such as rubber or oil (hydrogenous materials) in steel. Dynamic NR can also be used to monitor processes inside metallic containers; this being of interest in the fabrication of materials. The basics of the above techniques will be described along with examples in the areas of materials development and characterisation of relevance to advanced power systems.

1. INTRODUCTION

There is currently a great deal of debate internationally about the increased use of nuclear power in satisfying the ever-increasing demands for power throughout the world. While there are currently designs of power plant available that can provide power at a competitive cost and in great safety, there are many areas of these designs where significant improvements in efficiencies, and hence cost of power, could be made through the development and application of modern materials.

2. NEUTRON SCATTERING

Materials of interest to future power production systems are likely to be significantly different from those currently available. While it is possible to push operating temperatures from those currently available, significant increases, to 1000°C for example that has been mooted for several advanced systems, will require different approaches to both the fabrication of the materials themselves but also in the means used to join them into piping systems for example. It is also clear that information will be required on the resistance to radiation of the materials destined for in-core applications.

The neutron is an ideal tool for probing solids and liquids. Like light, electrons and X rays, neutrons can be used to reveal atomic structure. The way in which the neutrons are scattered provides information on the structure of the sample and the molecular dynamics in detail. Unlike protons and electrons the neutron has no charge, so it passes through the electron cloud which surrounds the nucleus. This makes neutrons very penetrative compared with electrons and X rays. It also means they can provide unique information about many elements, especially hydrogen, and isotopes of the same element.

Additional advantages of using neutrons are; that they have magnetic moments and so can probe magnetic materials, like those containing iron, to reveal magnetic structure, and that they have energies similar to the vibrational energy of atoms in solids and liquids. This means neutrons can reveal the motions of atoms in molecules.

2.1 Engineering

In-situ neutron studies have been used to check and improve the performance of welds in industrial components. This leads to improvements in both the mechanical and thermal performance of the components. Neutron imaging techniques are routinely used in quality control studies to detect internal flaws in critical equipment such as aircraft engine turbine blades. The aerospace, marine, petrochemical and defence industries benefit from these techniques.

Neutron scattering is also being used to determine the relationship between the composition and structure of materials, and their properties. This can provide information applicable in the design of new materials, such as opto-electronics, room temperature superconductors, nanostructures, and a new generation of tougher ceramics.

2.2 Earth and Environmental Sciences

Neutrons have been used to study minerals at high pressure and temperature to understand the geological history of earth. They can also provide much information on the shape and behaviour of minerals and on mineral extraction processes. This knowledge can assist in the development of more efficient mining practices. Another research area is the development of improved methods for the disposal of radioactive waste.

2.3 Physics

Physics is the starting point for understanding the fundamental properties of advanced materials, including magnetism, and superconductivity. Neutrons lend themselves to the study of dynamic phenomena such as superfluids and how the structure of materials change when the surrounding environment is varied; by, for example, temperature and pressure changes.

2.4 Chemistry

Neutron studies reveal crystal structure and dynamics of materials. They shed light on hydrogen bonding, how chemical reactions work and how atoms move in molecules. This increased understanding is of important benefit to the study of plastics, batteries and new materials that will be important in the development of hydrogen fuel systems, for example.

3. INSTRUMENT TYPES AND APPLICATIONS

There are a number of different neutron diffraction/scattering instruments that are available at research reactor and spallation neutron source facilities. Listed below are the main types and how they can be applied to the structural analysis of materials and components.

3.1 High-Resolution Powder Diffractometer

As research expands into new materials, the complexity of the structural problems increase. In these cases, a high-resolution instrument can more accurately determine the atomic and magnetic structures. This is because better resolution allows greater peak separation, hence greater accuracy in the determination of the peak intensities needed for structural refinement. There is a plethora of uses for high-resolution powder diffraction, some of which are listed here:

- The determination of crystal structures, ab initio, from powder diffraction data, such as when no suitable single crystal can be obtained.

- The examination of materials with complex crystal structures, including catalysts, incommensurate structures, hybrid materials, organics, cements, natural minerals, zeolites, phase separation, and non-linear optical materials. In many cases, the subtleties of the structure relate fundamentally to the material property.

- The study of materials that undergo phase transitions where small peak splitting is the key indicator of structural change. This includes areas of research into ferroic materials (ferroelectric, ferroelastic, ferromagnetic, etc), critical phenomena, and electronic materials such as superconductors and magneto-resistive materials.

- Monitoring of changes to the peak shape, which are good measures of strain, crystallite size, and the presence of defects. These are important parameters in materials such as batteries, hydrogen storage materials and mesoscopic structures. The coupled use of small angle neutron scattering may provide more insight into such materials.

3.2   High-Intensity Powder Diffractometer

The properties of a material are intrinsically linked to its atomic structure and hence the evolution of the material properties with temperature, pressure, applied fields and time mirrors the way in which the atomic structure evolves with these parameters. Therefore in order to understand the evolution of these properties in-situ parametric measurements of the atomic structure as a function of variations in these parameters need to be made. This is the scientific mission of HIPD instruments, to provide a vehicle for carrying out such in-situ parametric measurements. Some examples of such measurements are:

- Phase transitions; by varying one or more of the temperature, applied field (magnetic or electric) and applied pressure we can either create or destroy properties in a material. Typical examples are superconductivity, ferromagnetism, anti-ferromagnetism, ferroelectricity, piezo-electricity.

- Material formation; many materials undergo one or more chemical reactions as a function of time as they are formed. Some examples would be the drying of cement, solid state reactions and synthesis, growth from the melt. Real-time studies of these irreversible processes reveal the evolution of the structure at each stage.

- Cyclic variations; when materials are periodically (cyclically) exposed to external fields, for example, the application of electric or magnetic fields to ferroelectric or ferromagnetic materials, the resulting changes in the atomic structure can be followed in real time during the cyclic process.

3.3   Neutron Reflectometer

Neutron reflectometry is used to probe the structure of surfaces, thin-films or buried interfaces and processes occurring at surfaces and interfaces such as adsorption, corrosion, adhesion and interdiffusion. It is sensitive to structures on the nano scale (1 nm – 100 nm) in reflection mode and requires very good surfaces (usually thin films on silicon wafers, quartz, salt or glass for example). In recent years there has been a very rapid increase in interest in the biosciences as well as the emerging field of nanotechnology.

Experiments that the technique can be applied to include: specular scattering from air/solid, solid/liquid and in particular 'free liquid' samples, 'off-specular' scattering from the previous sample types and measurement of the kinetics phenomena on a minute or slower time scale. Also some interest has been expressed in the ability to conduct glancing-angle and wide-angle scattering studies for the investigation of short length scale, in-plane structures.

3.4 Triple-Axis Spectrometer

Neutron Inelastic Scattering is a widely used technique in condensed-matter physics. It is a key technique for the measurement of excitations in materials - collective excitations such as phonons and magnons, diffusive excitations like spin fluctuations and localised excitations arising from the hopping of charge in materials, crystal-field levels and some intramolecular modes.

While measurements of the structure of materials yield interatomic separations, measurements of the structural excitations in materials yield interatomic forces and measurements of magnetic excitations yield the force between magnetic spins. Neutron inelastic scattering is a key technique for extracting fundamental information about materials and this knowledge has been of great importance in understanding phenomena such as superconductivity and in the development of advanced materials.

3.5 Quasi-Laue Diffractometer

Single-crystal diffraction is a widely used technique to study structural properties of materials. This is mostly done using X rays and neutron single-crystal diffraction is less commonly used because of the special facilities needed. However, neutron single-crystal diffraction can be essential to study certain features, which cannot be studied using other methods and provides data that is complementary to that obtained using X rays.

The recent rapid expansion of topical areas of chemistry, such as supramolecular chemistry, crystal engineering and molecular modelling, requires accurate fundamental information concerning weak intermolecular interactions involving hydrogen atoms. Neutron diffraction provides the means for full characterisation of such interactions in the solid state.

Types of materials to be studied by single-crystal neutron diffraction in the context of materials research needs include:

- Organic compounds, including biologically important species
- Inorganic solids, organometallic complexes and coordination compounds
- Framework solids
- Metals and alloys
- Minerals
- Biological and synthetic polymer fibres
- Incommensurate solids
- Supramolecular structures

3.6 Small-Angle Neutron Scattering

Small-angle neutron scattering (SANS) is a key tool in the study of a variety of phenomena at the nanoscale, probing the structure of materials on a length scale from about one to several hundreds of nanometres. The size range spans a vast range of science, from proteins and viruses (biology and medical sciences) to emulsions and microemulsions (polymer and materials science) to phase separation and fractal growth (physics, geology and metallurgy).

Where possible, complementary imaging techniques such as transmission and scanning electron microscopy (TEM and SEM) should be used in conjunction with SANS to obtain the best results. However in many cases SANS is either the best or the only experimental technique available that is capable of providing structural and kinetic information concerning nano-sized inhomogeneities in the medium of interest, whether that be a matrix, a solution or one or more components in a mixture. For example, SANS does not require the production of very thin specimens as is the case for TEM; this process can destroy the very region of the material that is of interest. Microscopy is also less attractive for aqueous samples. Further, a microscopy image may include artefacts and may not be truly representative of the sample. While SANS does not provide real-space structure directly, the technique does probe the sample in its entirety.

The major strength of the SANS technique is that it can be used as a probe on a host of materials, which cover a wide range of research disciplines. Materials that are routinely characterised using the SANS technique include, alloys and ceramics, biological materials, colloidal materials, complex fluids, polymers, surfaces and interfaces and flux lattices in superconductors. Each of these areas has existing or potential industrial interest.

3.7　Polarisation-Analysis Spectrometer

In these instruments the neutron beam is pulsed using a chopper disc, allowing the energies of the neutrons to be measured simply by their arrival time at the detector. The technique is more applicable to soft matter and chemistry studies, including polycrystalline, glassy and liquid samples.

4.　RESIDUAL STRESS MEASUREMENT

An important aspect of structural fabrication using welding techniques is a knowledge of the stresses that build up as the material cools from the welding temperature. These so-called residual stresses add to the applied stresses and can be as high as yield stress levels. A variety of methods have been employed to measure these stresses; destructive techniques such as hole drilling, and non-destructive tests such as Barkhausen noise, ultrasonics and X ray diffraction. All these are either surface techniques or provide an average of the stress through the sample.

Neutrons have the advantage of being able to penetrate many tens of millimetres through materials of engineering significance, thus providing through-thickness measurement of residual stresses. The sampling volume can be made very small, down to about a 1 mm cube (at the expense of increased data collection times). The technique has been widely used for many years in the energy, aerospace, transport, petrochemical and general industries. Early machines (essentially first generation) relied on beam ports with relatively low neutron flux, and often many hours to obtain sufficient neutrons to characterise a single point. The advent of more powerful research reactors, and improvements in detectors allowed this sampling time to be reduced by several orders of magnitude.

Sample size has been a significant problem in the past. Early instruments had limited mass and space capacity, limiting the size of component that could be examined. This limitation for neutron based residual stress measurement is quite significant and has resulted in many on-site measurements being performed using portable X ray equipment.

4.1　Technique

Residual stress measurements by neutrons are exclusively performed with neutrons from research reactors or spallation sources. The requirements include, a goniometer (on which the specimen is positioned relative to the beam), an incident beam collimated by slits, detector slits (that act to provide the mask for defining the scanning volume) and a detector (or detectors) that enable the scattered neutrons to be counted. The arrangement is shown in Fig.1. Here the gauge volume is determined by the width of the slits. Measurements of residual stress are determined from the measurements of the lattice spacing within the material. This requires a sample that is known to be stress-free, and is normally provided by a sample made from small cubes (often only 1 mm side) glued together to give a sufficient volume to obtain good counting statistics.

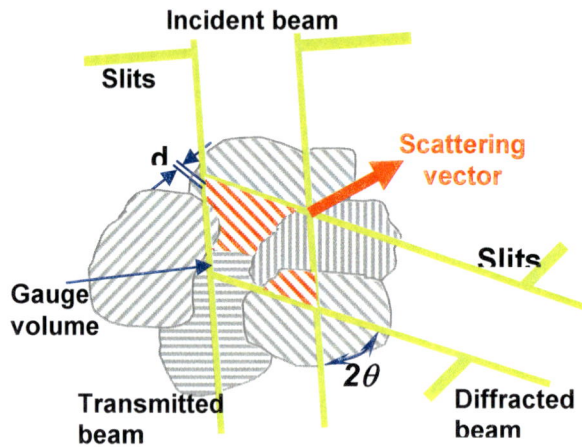

Fig 1 Schematic of the experimental arrangement for residual stress measurement.

There are limitations for the various techniques in respect to resolution and samples size. Figure 2 shows the current state of play for various inspection techniques. Neutron diffraction and residual stress measurements provide an important penetration depth range unmatched by many other techniques.

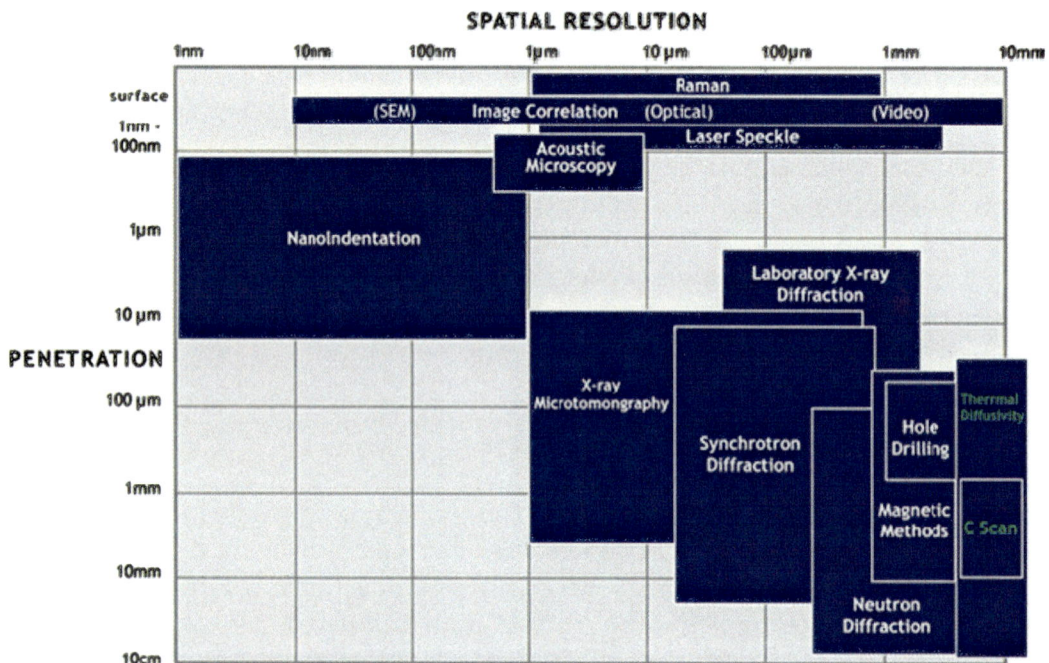

Fig. 2 Penetration depth of several inspection techniques and respective spatial resolution.

4.2   Example

The following is an example of what can be achieved with residual stress methods. It is related to the build-up of stress in quenched and tempered steel. Figure 3 shows the stress as a function of position across a weld. It indicates that stresses of yield magnitude can develop in joints which are highly restrained. Such stresses are detrimental to the structure as they can result in effects such as stress corrosion cracking in the absence of an applied load. Residual stress measurements are used to measure the stress in the sample after appropriate stress-relief processes, and have been used in a number of cases in weld procedure development.

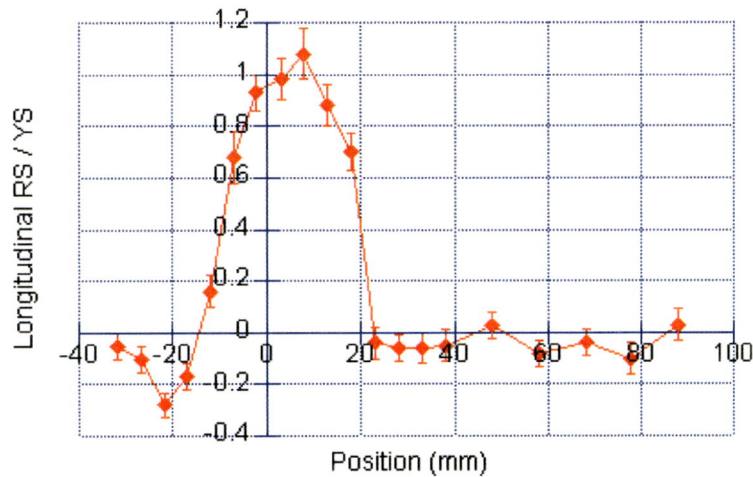

Fig 3 A graph showing longitudinal residual stress (as measured by neutron diffraction) as a function of position across a weld. The stress is zero in the bulk of the material and rises to yield stress in the weld.

5. NEUTRON RADIOGRAPHY

As has been noted above, neutrons possess the ability to pass through many tens of millimetres of materials of engineering significance. This has resulted in them being used for radiography. Neutrons have an advantage over the conventional radiography modalities of X and gamma rays, of not being sensitive to atomic weight in terms of absorption. Elements such as hydrogen, cadmium, boron and gadolinium all absorb neutrons very strongly, while a number of common engineering materials, such as aluminium and steel, absorb much less strongly.

Hydrogen is also an element that strongly absorbs neutrons. This fact leads to the ability of neutron radiography to detect light, hydrogenous material within high density material such as steel. This has been used in many instances for detecting mis-placed "O" rings in cylinders or for determining the level of explosive powder in ammunition cases.

5.1 Technique

The technique of neutron radiography is very similar to those used for conventional radiography with a few significant exceptions. Whereas conventional radiography attempts to obtain the best images (sharpest) through the use of small sources, or by using a large sample-to-film plane distance, this is not feasible with neutrons; high intensity point sources not being available. In order to minimise geometric unsharpness, neutron radiographs have historically been taken with the film very close to, or in contact with the sample. In addition the film is not exposed by the neutrons, but rather by beta particles from a thin sheet of gadolinium foil that is kept in intimate contact with the film. The neutrons pass through the film are absorbed by the gadolinium, which then emits a beta particle, thus exposing the film in the immediate area. Film radiography is now being replaced with image plates, providing digital capture of the image and the inherent advantages of recording in this manner; namely, the ease of long term storage and the ability to manipulate the images to improved contrast etc.

There are essentially three methods of obtaining images in neutron radiography. The first is by the indirect or transfer method. In this method a material, such as indium or dysprosium, is placed in the beam that is activated by the neutrons. This is then removed from the beam and is placed on a sheet of film. This then exposes the film and provides the radiograph. This method is useful where there is significant gamma contamination of the beam or where the item being inspected is radioactive.

The direct method is widely used in inspecting components such as turbine blades or similar items. In this method the film is placed in the beam, but is exposed by the beta decay of a sheet of thin material such as gadolinium.

A third method, called track-etch imaging, is useful in the case where the item being examined emits heavy ions after absorbing neutrons. A plastic film is positioned close to the item and the ions produce damage in the film. Subsequent etching of the material (using for example potassium hydroxide) will reveal damage pits, as the ion damaged areas are more easily etched.

5.2 Examples

There are many examples of neutron radiography available, a number of which are indicative of the use to which the technique can be put to assist in the development of advanced materials. One of the most common components examined are turbine blades for aircraft engines or power generators. Examples can be seen at the web sites of organisation such as the Paul Scherrer Institute (http:/neutra.web.psi.ch/), CEA (http://www-llb.cea.fr/neutrono/nr1.html) and Nray Services Inc (http://www.nray.com/).

6. FACILITIES AROUND THE WORLD

Following are lists of facilities that provide neutron based research techniques. Both reactor and spallation sources are included; and an attempt has also been made to include commercial facilities where appropriate, particularly in the area of neutron radiography. It is beyond the scope of this paper to identify the types of instruments at each facility. This can usually be achieved by referring to the web sites of the facilities individually (the ANL web site has links).

6.1 Neutron scattering facilities *(source http://www.neutron.anl.gov/facilities.html)*

6.2 Asia and Australia

- Bragg Institute, Australian Nuclear Science and Technology Organisation, Lucas Heights, Australia

- High-flux Advanced Neutron Application Reactor (HANARO), Korea

- Japan Atomic Energy Research Institute (JAERI), Tokai, Japan

- KENS Neutron Scattering Facility, KEK, Tsukuba, Japan

- Kyoto University Research Reactor Institute (KURRI), Kyoto, Japan

- Malaysian Institute for Nuclear Technology Research (MINT), Malaysia

6.3 Europe

- Budapest Neutron Centre, AEKI, Budapest, Hungary

- Berlin Neutron Scattering Center, Hahn-Meitner-Institut, Berlin, Germany

- Center for Fundamental and Applied Neutron Research (CFANR), Rez nr Prague, Czech Republic

- Frank Laboratory of Neutron Physics, Joint Institute of Nuclear Research, Dubna, Russia
- FRJ-2 Reactor, Forschungzentrum Jülich, Germany
- FRM-II Research Reactor, Garching, Germany
- GKSS Research Center, Geesthacht, Germany
- Institut Laue Langevin, Grenoble, France
- Interfacultair Reactor Instituut, Delft University of Technology, Netherlands
- ISIS Pulsed Neutron and Muon Facility, Rutherford-Appleton Laboratory, Oxfordshire, UK
- JEEP-II Reactor, IFE, Kjeller, Norway
- Laboratoire Léon Brillouin, Saclay, France
- Ljubljana TRIGA MARK II Research Reactor, J. Stefan Institute, Slovenia
- Risø National Laboratory, Denmark
- St. Petersburg Nuclear Physics Institute, Gatchina, Russia
- Studsvik Neutron Research Laboratory (NFL), Studsvik, Sweden
- Swiss Spallation Neutron Source (SINQ), Villigen Switzerland

6.4 North and South America

- Centro Atomico Bariloche, Rio Negro, Argentina
- Chalk River Neutron Programme for Material Research, Chalk River, Ontario, Canada
- High Flux Isotope Reactor (HFIR), Oak Ridge National Laboratory, Tennessee, USA
- Indiana University Cyclotron Facility (IUCF), Bloomington, Indiana, USA
- Intense Pulsed Neutron Source (IPNS), Argonne National Laboratory, Illinois, USA
- Los Alamos Neutron Science Center (LANSCE), New Mexico, USA
- McMaster Nuclear Reactor, Hamilton, Ontario, Canada
- MIT Nuclear Reactor Laboratory, Massachusetts, USA
- NIST Center for Neutron Research, Gaithersburg, Maryland, USA
- Peruvian Institute of Nuclear Energy (IPEN), Lima, Peru
- University of Missouri Research Reactor, Columbia, Missouri, USA
- University of Illinois Triga Reactor, Urbana-Champaign, Illinois, USA

6.5 New Projects

- Australian Replacement Research Reactor, Lucas Heights, Australia
- Austron Spallation Neutron Source, Vienna, Austria
- Canadian Neutron Facility, Chalk River, Ontario, Canada
- European Spallation Source (ESS)

- Japan Proton Accelerator Research Complex (J-PARC), Tokai, Japan
- Low Energy Neutron Source (LENS), Indiana University Cyclotron Facility, USA
- Spallation Neutron Source, Oak Ridge National Laboratory, Tennessee, USA

6.6   Residual Stress Facilities

The number of residual stress facilities is limited. The list below does not include the new instrument being developed at the OPAL reactor in Australia.

- NIST, USA
- ORNL, USA
- ANSTO, Australia
- ISIS, UK (spallation source)
- Hahn Meitner Institute, Germany
- BNSC, Berlin Neutron Scattering Centre, Germany
- NFL, Studsvik Neutron Research Laboratory, Sweden

6.7   Neutron Radiography Facilities

There are a number of facilities available in mainly Europe and North America providing neutron radiography services to research and industry. The following list gives the currently operating facilities.

- Paul Scherrer Institute, Switzerland
- DRN/DRE/SIREN-CEA/Saclay, France
- University of Texas at Austin, USA
- ATI, Vienna, Austria
- Del Mar Ventures, San Diego, USA (seller of pulsed neutron sources)
- LANSCE, USA
- MIT, USA
- Aerotest Operations, California, USA (Commercial service provider)
- NRay Services Inc, Ontario, Canada (Commercial service provider)
- Budapest Neutron Centre, Hungary
- Cornell University, USA
- Bhabha Atomic Research Cnntre, India
- Indira Gandhi Centre for Atomic Research, India

7.   ACKNOWLEDGEMENTS

The material in this paper came from a number of sources, principally from the Bragg Institute website associated with ANSTO. The author would also like to acknowledge the contribution of Dr Shane Kennedy of the Bragg Institute for his input to the use of the diverse range of neutron scattering instruments.

Materials characterization using positron annihilation spectroscopy

G. AMARENDRA, R. RAJARAMAN, C. S. SUNDAR, B. RAJ

Indira Gandhi Centre for Atomic Research,, Kalpakkam , India

Abstract: An overview of core structural materials for thermal and fast reactors is presented in terms of various microscopic and macroscopic changes that take place due to intense neutron irradiation and consequent radiation damage. Various experimental techniques for materials characterization in terms of structural, microstructural and chemical properties are briefly discussed. Positron Annihilation Spectroscopy (PAS) consisting of positron lifetime, Doppler broadening and Angular correlation techniques are illustrated. The application of PAS for investigating helium bubble growth through (n,α) reaction using a thermal reactor and direct helium injection using a cyclotron, is highlighted in the case Cu and the extension of these results is discussed for Ni and Stainless steels. Depth-resolved non-destructive probing of near-surface defects in irradiated steel alloys is discussed.

1. INTRODUCTION

The physical and chemical properties of structural materials used in the core components of nuclear fission components undergo drastic changes owing to intense neutron environment and consequent radiation damage phenomena [1-4]. Figure 1 schematically shows this interplay between radiation and changes in physical and chemical properties.

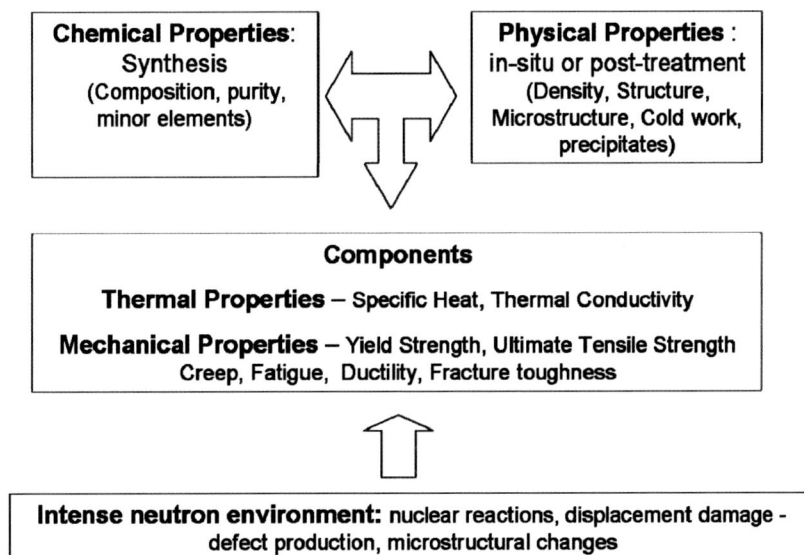

Fig. 1 Structural materials are chosen based on their physical and chemical properties, so as to yield desirable thermal and mechanical behavior. Due to intense neutron background these properties get deleteriously altered.

Production of various atomistic defects such as vacancies, interstitials and their diffusion and agglomeration to form higher order defects lead to microstructural changes, which have macroscopic consequences as depicted in Fig. 2. The introduction of helium due to (n,α) reaction in the matrix gives rise to stabilization of voids as gas bubbles. This leads to the void swelling problem, which is of serious consequence in high helium production environment like fast and fusion reactors.

As shown in Fig. 2, the helium to vacancy ratio and the temperature regime essentially determine void / helium bubble nucleation and growth behaviour. Table 1 depicts some of the important issues arising due to the radiation effects on the structural materials. While thermal reactors suffer mostly from hydrogen embrittlement, swelling is the limiting issue in the case of fast breeder reactors.

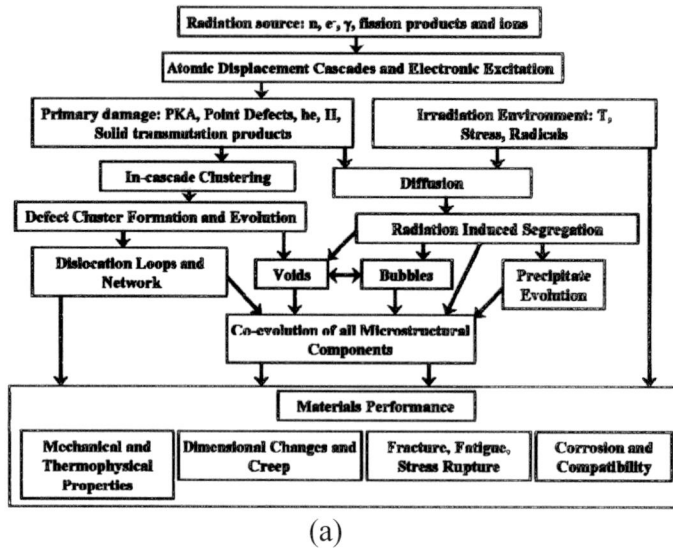

(a)

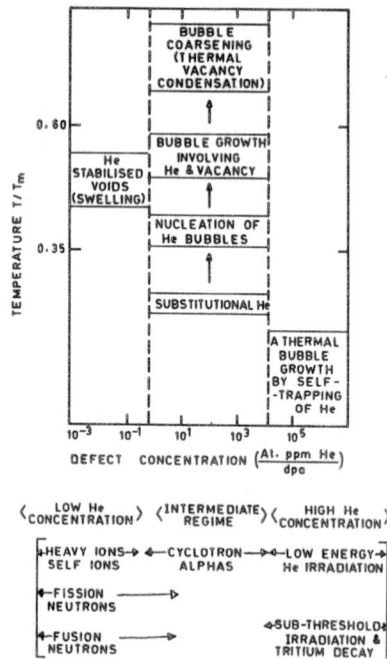

(b)

Fig. 2 Microscopic and macroscopic consequences of radiation damage in structural materials. The figure (a) depicts various stages of defect production and consequent macroscopic changes, while the figure (b) shows various issue related helium production, bubble formation and growth for various helium production rates and temperature regimes.

Table 2 lists various techniques used in characterization of structural and chemical aspects of materials [5]. Figure 3 sketches different kinds of defects ranging from point defects such as vacancy, interstitial, impurity atoms, line defects like dislocations and dislocation loops and three dimensional defects like voids and precipitates.

Among various techniques employed in studying defects in metals and alloys, as listed in table 3, Transmission Electron Microscopy (TEM), Small Angle Neutron Scattering (SANS) and Positron Annihilation Spectroscopy (PAS) are indispensable tools. In this paper, we will review the application of positron annihilation technique to radiation damage studies with selected examples mostly from our work.

TABLE 1. SOME OF THE IMPORTANT MATERIALS AND THEIR PROBLEMS IN THERMAL AND FAST REACTORS

Reactor type	Materials	Problems
Thermal Reactors	Ziraloy-2 and 4, Al alloys, Ferritic Steel	Hydrogen Embrittlement. Irradiation Creep, Corrosion, Fracture toughness-NDTT, Transmutation products
Fast Breeder Reactors	Austenitic Steels-SS304, SS316, SS316-LN, Ti-stabilized SS316, D9 Steel (15%Ni, 14%Cr)	Void swelling, Irradiation creep, Fatigue, Microstructural changes

TABLE 2. AN OVERVIEW OF SOME OF THE STRUCTURAL AND CHEMICAL CHARACTERIZATION TECHNIQUES, LISTED UNDER VARIOUS CATEGORIES [5]

Structural Characterization	Chemical Characterization
Bulk structure – XRD, TEM, HRTEM, Neutron scattering	**Bulk composition** – Spectrophotometry, Mass spectroscopy
Surface structure – GIXRD, FIM, LEED, Ion Channeling, STM	**Surface composition** – AES, XPS, SIMS, EPMA, RBS
Surface topography – Optical Microscopy, SEM, AFM, Ellipsometry	
Microstructure *(Impurity states, Interstitials, Vacancies, Color centres, dislocations, Grain boundaries)* – HRTEM, UVVIS, Raman, PL, FTIR, MS, PAC, PAS **Microstructure** *(Cracks, Voids, Precipitates, Exclusions)* – NDT, TEM, SAXS, SANS	**Impurities & local chemical environment** – Mass Spectroscopy, ESR, NMR, Mass spectroscopy, EPMA, AES, XPS, SIMS, PIXE, Raman, PL, FTIR, PAC, PAS, XRF, EXAFS
Magnetic Structure – Neutron scattering, MS, PAC	
XRD- X ray Diffraction; TEM- Transmission Electron Microscopy; HRTEM- High Resolution TEM; GIXRD- Glancing Incidence XRD; FIM- Field Emission Microscopy; LEED- Low Energy Electron Diffraction; STM- Scanning Tunneling Microscopy; SEM- Scanning Electron Microscopy; AFM- Atomic Force Microscopy; UVVIS- Ultra Violet Visible optical Absorption Spectroscopy; PL- Photo Luminescence; FTIR- Fourier Transform Infrared Spectroscopy; MS- Mossbauer Spectroscopy; PAC- Perturbed Angular Correlation; PAS- Positron Annihilation Spectroscopy; NDT- Non Destructive Testing; SAXS- Small Angle X ray Scattering; SANS- Small Angle Neutron Scattering; AES- Auger Electron Spectroscopy; XPS: X ray Photoemission Spectroscopy; SIMS: Secondary Ion Mass Spectrometry; EPMA- Electron Probe Micro Analysis; RBS- Rutherford Back Scattering; ESR- Electron Spin Resonance; NMR- Nuclear Magnetic Resonance; PIXE- Particle Induced X ray Emission; XRF- X ray Fluorescence; EXAFS- Extended X ray Absorption Fine Structure.	

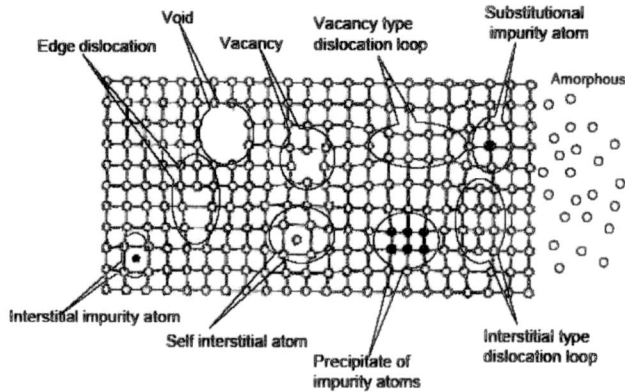

Fig. 3 Schematic of different atomistic defects in solids, whose production, diffusion and segregation leads to deleterious macroscopic consequences

TABLE 3. SELECTED LIST OF VARIOUS EXPERIMENTAL TECHNIQUES FOR PROBING DEFECTS AT MICROSCOPIC AND MACROSCOPIC DIMENSIONS IN METALS, ALLOYS AND SEMICONDUCTORS.

Metals & Alloys	Semiconductors
• Resistivity • Dilatometry • Nondestructive testing (NDT) • XRD, SAXS • Electron Microscopy (SEM, TEM) • Small Angle Neutron Scattering (SANS) • Positron annihilation • Perturbed Angular Correlation	• Resistivity • Dilatometry • NDT • XRD, SAXS • Electron Microscopy • SANS • RBS, Ion Channeling • EPR, NMR • Mossbauer Spectroscopy • Positron Annihilation • Perturbed Angular Correlation • Raman Scattering • Infrared & Optical absorption • Ellipsometry

2.   POSITRON ANNIHILATION SPECTROSCOPY

Positron, the antiparticle of electron, will annihilate when it encounters an electron in its close proximity in the medium, as depicted in figure 4. This mass to energy conversion process mostly leads emission of two 511 keV γ-rays emitted in opposite directions to conserve momentum and energy. A single γ emission requires a third body to take care of recoil momentum, which eliminates this mode under normal conditions. On the other hand, 3γ emission is possible with negligible probability. The conservation of momentum and energy of annihilating electron positron pair provide information on electron momentum, whereas the annihilation rate signifies the average electron density in the medium. Figure 5 shows schematic representations of three types of positron annihilation measurements viz., positron lifetime, angular correlation and Doppler broadening [6,7]. When energetic positrons, obtained from radioactive isotopes like ^{22}Na, are injected into sample they first slow down to finally reach thermal energy. The range of positron implantation in metals is of the order of few tens of micrometer. The thermalised positron randomly diffuses in the medium before

28

annihilating with an electron. The typical diffusion length is ~100 to 400 nm in crystalline solids.

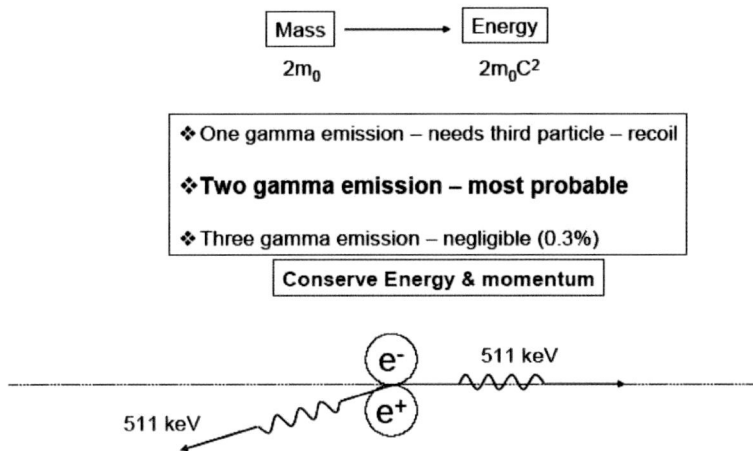

Fig. 4 Positron annihilation with an electron of the medium in metals and alloys results in the emission of two 511 keV gamma rays in opposite directions, by probing which information pertaining to defects can be obtained.

The annihilation event produces two 511 keV γ-rays emitted in opposite direction to conserve momentum and energy. Since the momentum of thermalised positron is negligible, the net momentum is essentially that of electron. In the laboratory frame the transverse components of electron momentum result in angular deviation between emitted γ-rays from co-linearity. The longitudinal component Doppler shifts 511 keV γ-ray energy.

While the angular correlation can be measured with very high resolution, the state of the art energy resolution of γ-ray spectrometry is of the same order of the Doppler shift obtained for typical electron momentum. From Doppler broadening measurements, one can obtain qualitative comparison of various defect states by defining lineshape parameters viz., S-parameter signifying the fraction of positrons annihilating with valance electrons and W-parameter signifying the fraction of core electron contributions. The average time spent by the positron in the sample before annihilation is inversely proportional to the local electron density. The lifetime of positron, inverse of the annihilation rate, can be estimated by measuring time differences between a prompt γ-ray of energy 1.28 MeV and the 511 keV γ-ray resulting from the annihilation. In defect-free bulk samples, positron is in delocalized state sampling mostly the interstitial region due to the repulsion from ion cores. When open volume defects like vacancies are encountered, positron gets localized there due to less repulsion. The resulting annihilation rate will be lower as compared to defect-free state as the local electron density at defect will be lesser. Thus, positron lifetime increases due to the presence of open volume defects. Typical fractional defect concentration range of 10^{-7} to 10^{-4} is amenable for detection by positrons. This selectivity and sensitivity of positron to open volume defects make the positron annihilation technique as an important tool in the radiation damage studies. Since positrons emanating from radioactive isotopes have energies varying all the way from zero to endpoint energy, they will be sampling the whole depth region of a few tens of micrometer of sample. Thus, they cannot be used for studying defects in thin films and interfaces. However variable low energy positron beams can be produced for such specific depth resolved studies by moderating the positrons emitted from radioactive isotopes with materials having negative work function such as W(100) [8].

In this paper, we will demonstrate the capabilities of PAS technique with selected examples drawn from our work [9-16].

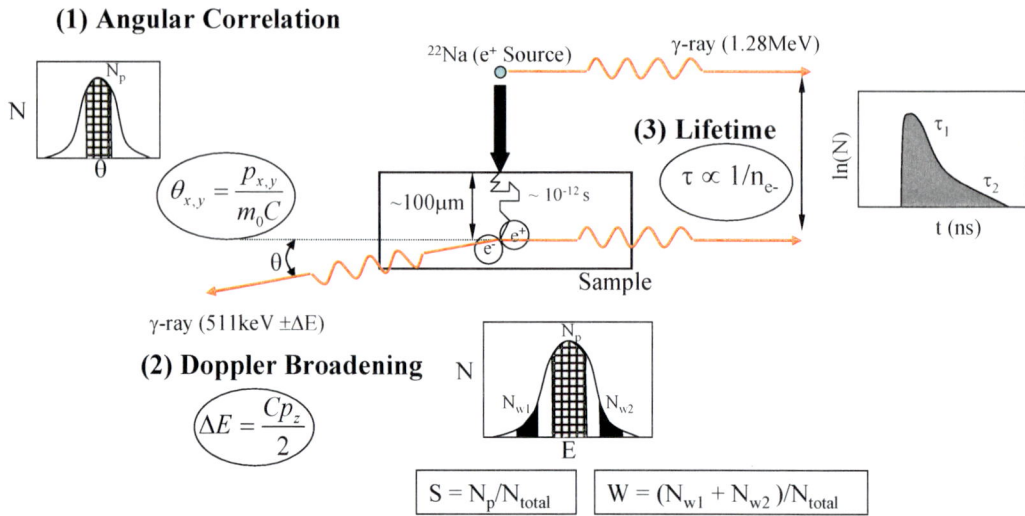

Fig. 5 Positron annihilation spectroscopy comprises of Angular correlation, Doppler broadening and positron lifetime techniques. For defects studies, the latter two techniques are extensively used.

3.   RESULTS FROM PAS

Figure 6 shows the positron lifetime results on electron irradiated Ni subjected to isochronal annealing treatment [9]. In the as-irradiated state mostly monovacancies were present and show typical lifetime of about 175 ps as compared to defect-free nickel lifetime of 110 ps. When the annealing temperature is increased further, these monovacancies migrate and form clusters resulting in higher positron lifetime. The average cluster size can be estimated by calculating positron lifetime for various defect cluster configurations as shown in Fig. 6(c), wherein the lifetime variation is plotted as a function of number of vacancies in the cluster. As seen from the figure 6(b), theoretical estimates compare well with experimental lifetimes, enabling quantitative estimates of cluster size and concentration from positron lifetime experiments.

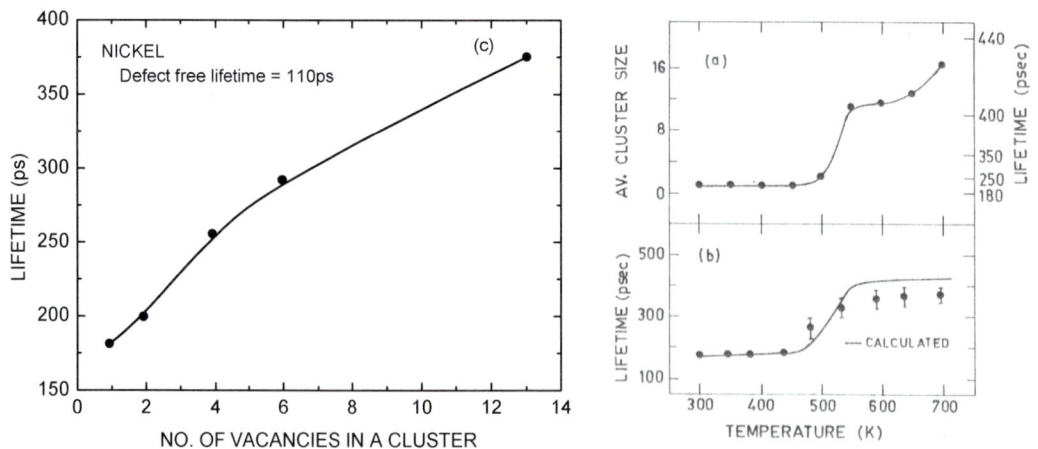

Fig. 6 Sensitivity of positron lifetime to vacancy cluster size is depicted based on experimental and theoretical work on cold-worked Ni.

Figure 7 shows positron lifetime results on helium bubble formation and growth with helium produced by nuclear transmutation in a reactor and by high energy alpha particle implantation in copper [10-11]. The characteristic feature of the isochronal annealing studies is the observation of a minimum for defect lifetime concurrent with a maximum in corresponding intensity at intermediate temperatures. This temperature regime is identified to be the helium bubble nucleation stage. Beyond this temperature the helium bubbles start growing leading to reduction in defect intensity and increase in defect lifetime. From the measured positron lifetime and intensity corresponding to helium bubble growth regime, one can estimate the helium bubble parameters namely, helium atom density inside bubble, bubble radius and bubble density. Figure 8 shows the TEM micrograph of the n-irradiated and annealed Cu-B sample. The measured helium bubble size of 4.5 nm correlates well with that obtained from positron lifetime results. Apart from studies on copper, extensive positron studies have been carried out on helium bubble formation and growth in various metals and alloys upon alpha implantation [12,13].

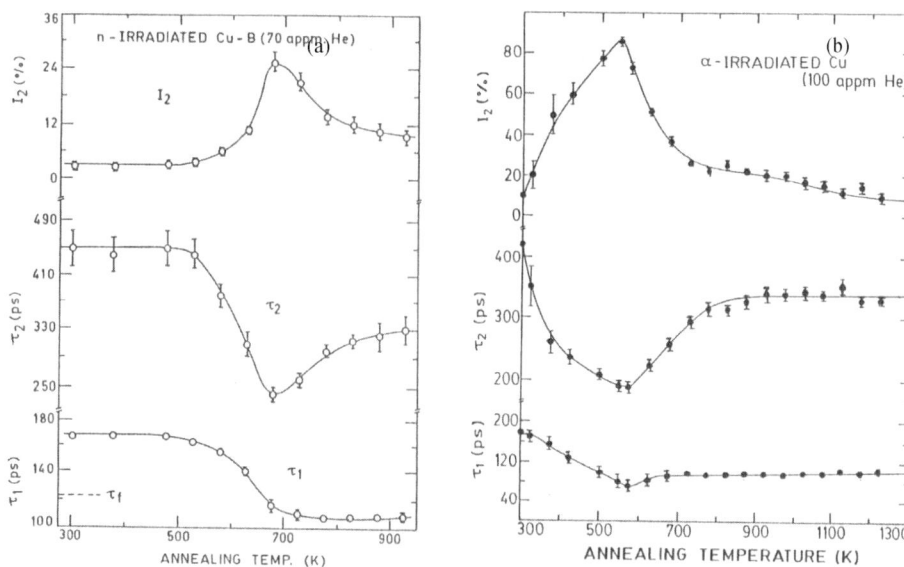

Fig. 7 Variation of positron lifetime parameters with annealing temperature for (a) n-irradiated Cu-B and (b) α-implanted Cu. Helium bubble growth at higher temperatures is not influenced by the mechanism of helium production.

Formation of intermetallic precipitates in β-quenched Zicaloy-2 samples was investigated by positron annihilation, hardness and ultrasonic velocity measurements [14]. Figure 9 shows variation of positron lifetime and Doppler line shape parameter as a function of annealing temperature in zircaloy-2. The positron annihilation parameters exhibit an increasing trend during 673–873 K temperature range, indicating the formation of misfit dislocations due to the occurrence of intermetallic phases. Complementary changes are observed in hardness, ultrasonic velocity and metallographic measurements. While positron annihilation studies provided information about atomistic interfacial defects associated with intermetallic precipitates, the ultrasonic velocity measurements were probing the changes in mean modulus. Hardness and metallography measurements provided macroscopic evidence towards the changes brought about by the intermetallic precipitates.

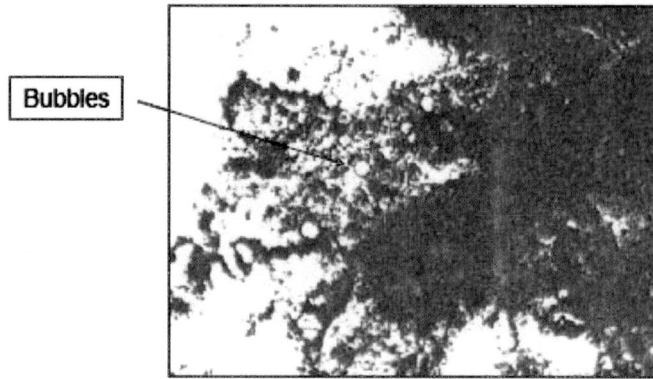

Fig. 8 Transmission electron micrograph of n-irradiated Cu-B revealing helium bubbles.

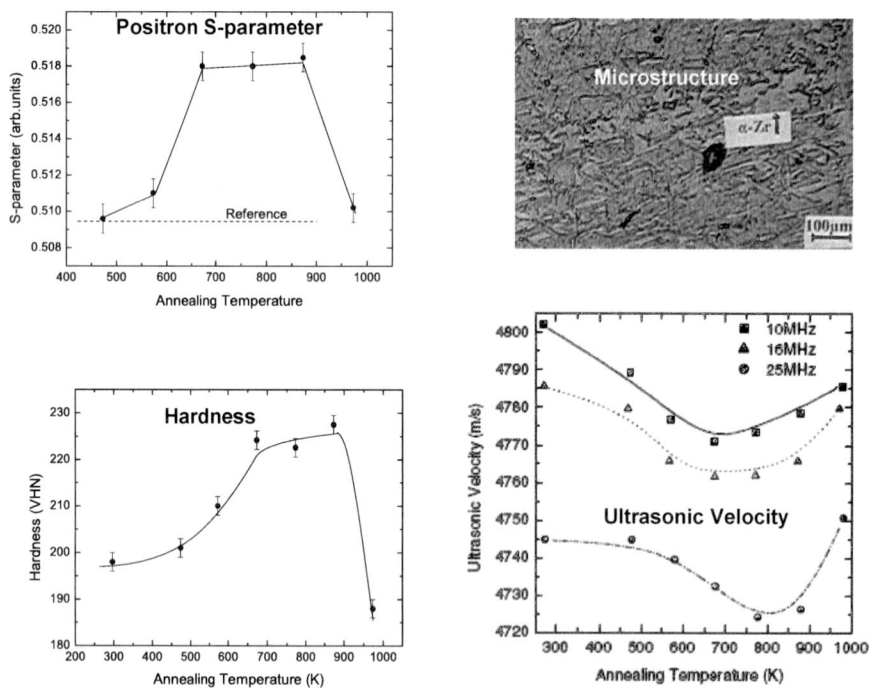

Fig. 9 Combined study of positron annihilation, microhardness, ultrasonic velocity measurements on intermetallic precipitation in zircaloy-2.

Figure 10 shows positron studies on TiC precipitation in Ti-stabilised austentic stainless steel (D9) [15]. Only one lifetime component is resolved in the entire annealing range. Positron lifetime in the solution annealed state for D9 alloy is 110 ± 1 ps and that in the 20% CW sample is 167 ± 2 ps, which corresponds to saturation trapping of positrons at cold work induced defects. The observed variation of lifetime τ in Fig. 10 shows four distinct stages viz., (a) an initial monotonic decrease in τ from CW state till 823 K, which corresponds to point defect recovery, (b) a sharp increase in the interval 823 - 973 K, which corresponds to the formation of TiC precipitates, giving rise to positron trapping at interfacial defects, (c) a plateau region in the interval of 973 - 1073 K, suggesting the completion of TiC precipitation and stability of these precipitates and (d) beyond 1073 K, a final decrease to near solution annealed state corresponding to the growth of TiC precipitates during recrystallisation.

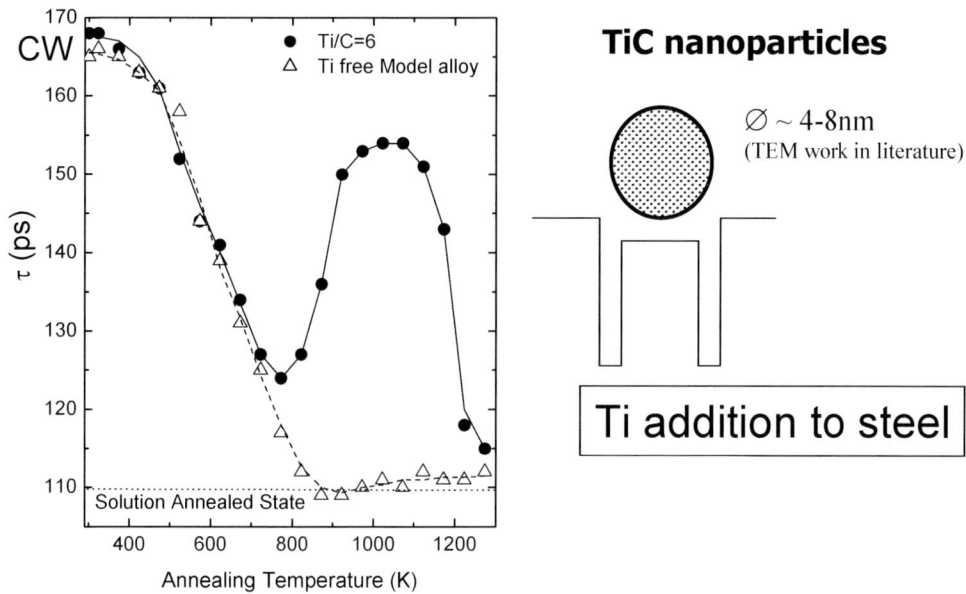

Fig. 10 Positron lifetime studies on TiC precipitation in Ti-modified stainless steel (D9). The open triangles correspond to Ti free reference alloy and filled circles correspond to D9 alloy. The pictorial representation shows the positron trapping potential at TiC precipitate-matrix interfaces.

From the variation of positron lifetime with annealing temperature for cold worked samples of Ti-free and Ti containing steel, the formation and growth of TiC precipitation stages are delineated. It is understood that the positrons get trapped at the interfacial defects at nanometer sized TiC precipitates.

Figure 11 demonstrates the application of variable energy positron beam to study the defects in thin films. Doppler lineshape parameter is monitored during depth profiling studies on defects in diamond films formed with Cr_2N or Si buffer layer on steel substrate [16]. From the variation of S-parameter with sample depth, it is observed that the Cr_2N buffer layer is ideal choice to form diamond film with lesser defects as compared to Si as a buffer.

Void swelling behaviour of a (15Ni-14Cr) Ti-modified steel, simulated by heavy ion irradiation, has been investigated using step height and positron annihilation measurements [17]. Figure 12 compares depth profiling studies of defects in Ni-irradiated SS316 and D9 steel to a dose of 84dpa. As seen from the figure the S-parameter shows lower value for D9 as compared to SS316 for the defected depth region. This is understood as follows. The fine TiC precipitates, which form in D9 at the irradiation temperature regime, are excellent sinks for point defects. This leads to less void formation and growth in D9 sample resulting in reduced swelling, as compared to SS316 sample.

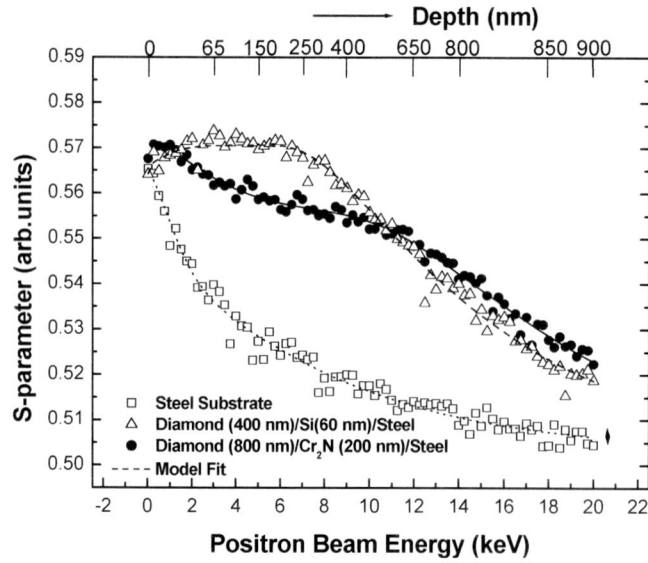

Fig. 11 Depth profiling of Diamond films coated on steel substrates with buffer layers using variable energy positron beam.

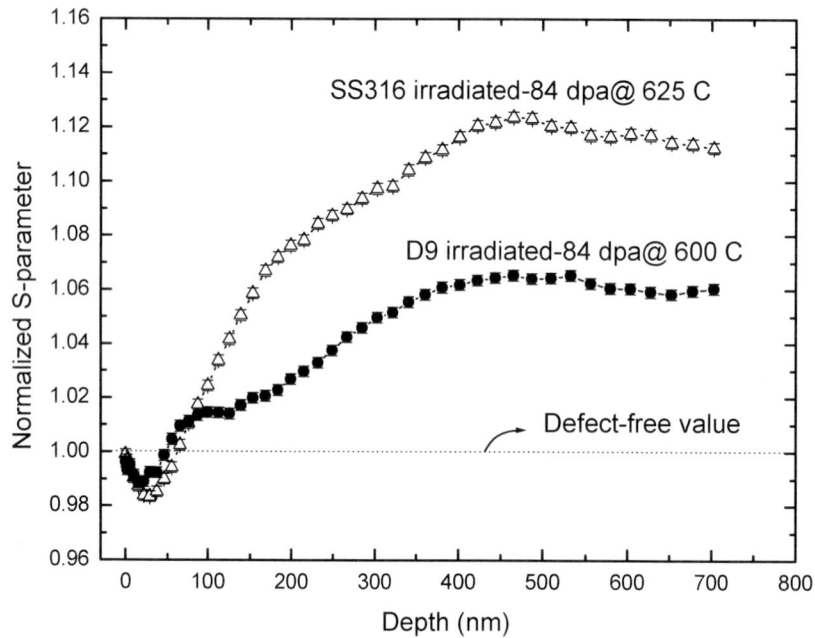

Fig. 12 Comparative depth profiling studies of defects in Ni-irradiated SS316 and D9 steel samples.

D9 samples have been 18% cold-worked, pre-implanted with helium at 170 keV and 275 keV energies so as to create a uniform helium concentration of 100 appm around 600 nm depth region [17]. This was followed by a 2.5 MeV Ni ion irradiation to create a peak damage of ~ 84 dpa at various irradiation temperatures between 700 and 970 K. Defect-sensitive positron annihilation lineshape S-parameter shows clear changes consequent to void formation as a function of sample depth as shown in fig. 13. The range of Ni ion is indicated by vertical dotted line in Fig. 13. From the variation of average S-parameter as a function of irradiation temperature, the peak swelling temperature has been deduced to be 870 K.

84 dpa, 100 appm He

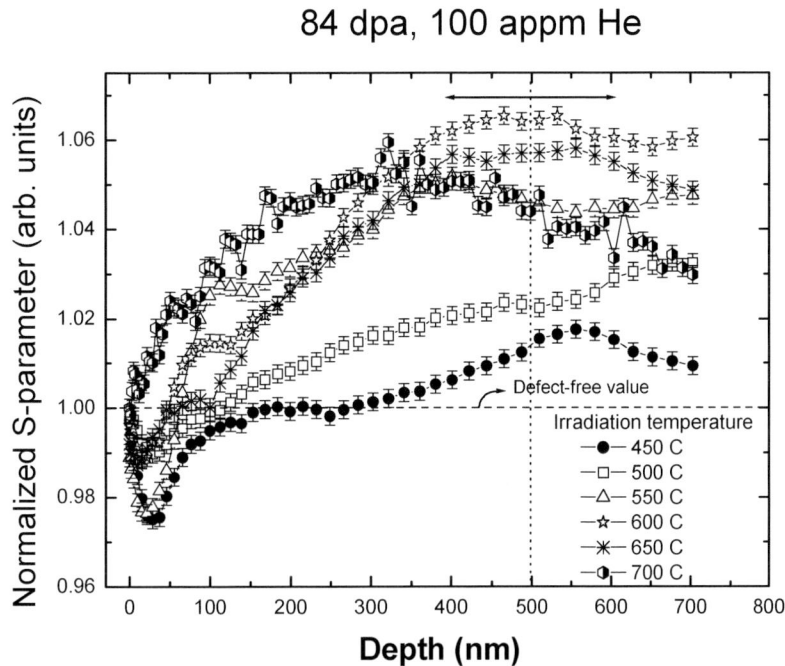

Fig. 13 Positron beam depth profiling studies of void formation in Ni-irradiated D9 steel

4.   SUMMARY

An overview of various microstructural changes and radiation damage in reactor structural materials is given. Positron Annihilation Spectroscopy (PAS) in terms of positron lifetime, Doppler broadening and Angular correlation techniques are illustrated. The application of PAS for comparing the helium bubble growth through (n,α) reaction using a thermal reactor and direct helium injection using a cyclotron, is highlighted in the case Cu and the extension of these results for Ni and Stainless steel presented. Depth-resolved positron beam studies on thin film coatings and irradiated alloys are highlighted.

REFERENCES

[1]   GARNER, F.A., Irradiation performance of cladding and structural steels in liquid metal reactors, , Chapter 6: "Irradiation Performance of Cladding and Structural Steels in Liquid Metal Reactors," Materials Science and Technology: A Comprehensive Treatment, VCH Publishers, Vol **10A** 1994, pp. 419-543.

[2]   SCHILLING, W., ULLMAIER, H., Physics of radiation damage in metals, , Chapter 9: "Physics of radiation damage in metals", Vol. 10B of Materials Science and Technology: A Comprehensive Treatment, VCH Publishers, 1994, pp. 180-241.

[3]   LEE, E.H., MANSUR, L.K., Fe-15Ni-13Cr austenitic stainless steels for fission and fusion reactor applications-II: Effects of minor elements on precipitate phase stability during thermal aging, , J. Nuclear Mater. **278** (2000) 11.

[4]   STOLLER, R.E., ZINKLE, S.J., NICHOLS, J.A., CORWIN, W.R., Report of Workshop on "Advanced Computational Materials Science: Application to Fusion and Generation IV Fission Reactors, by, Oak Ridge National Laboratory (2004) ORNL/TM-2004/132; see

http://www.csm.ornl.gov/meetings/SCNEworkshop/Workshop-Report-ORNL-TM-2004-132.pdf.

[5]   AMARENDRA, G., BALDEV RAJ, MANGHNAN M.H., (Eds.) Recent Advances in Materials Characterization, Universities Press, Hyderabad, CRC Press, USA (2007).

[6]   W. BRANDT, A. DUPASQUIER (Eds),Positron Solid State Physics, Amsterdam, North-Holland (1983)

[7]   WEST, R.N., Positron studies of condensed matter,., Adv. Phys. **22** (1973) 263.

[8]   SCHULTZ, P.J., LYNN, K.G., Interaction of positron beams with surfaces, thin films and surfaces, , Rev. Mod. Phys **60** (1988) 701.

[9]   SUNDAR, C.S., et al Positron annihilation studies of electron-irradiated, cold-worked and hydrogen-charged nickel, , Philos. Mag. A **50** (1984) 635.

[10]   VISWANATHAN, B., AMARENDRA, G., Helium bubbles in neutron-irradiated copper-boron studied by positron annihilation, , Radiation Effects, **107** (1989) 121.

[11[AMARENDRA,, G., VISWANATHAN, B., GOPINATHAN, K.P., Positron Annihilation study of helium clustering in alpha irradiated copper, Radiation Eff. and defects in Solids **118** (1991) 357.

[12]   AMARENDRA, G., Positron annihilation studies of helium in metals and alloys, Ph.D thesis, 1990, University of Madras, unpublished.

[13]   RAJARAMAN, R., Positron annihilation studies of light impurities in metals, Ph.D thesis, 1994, University of Madras, unpublished.

[14]   GOPALAN, P. et al, Characterisation of β-quenched and thermally aged Zircaloy-2 by positron annihilation, hardness and ultrasonic velocity measurements, , , Journal of Nuclear Materials **345** (2005) 162.

[15]   RAJARAMAN, R., PADMA GOPALAN, VISWANATHAN, B., VENKADESAN, S., A positron annihilation study of TiC precipitation in plastically deformed austenitic stainless steel, , J. Nucl. Mater. **217** (1994) 325.

[16]   SHANKAR, P., et al, Depth profiling of Diamond films with Cr_2N and Si buffer layers using variable energy positron beam, (unpublished)

[17]   CHRISTOPHER DAVID, et al, Void swelling studies in ion irradiated (15Ni-14Cr), Ti-modified stainless steel using positron annihilation and step height measurements, J. Nucl. Mater. (2006) submitted.

Current activities and future facilities in Canada

A. MCIVOR

National Research Council Canada, Canadian Neutron Beam Centre,
Chalk River Laboratories, Chalk River, Canada.

Abstract: The capabilities of the NRC Canadian Neutron Beam Centre, its suite of spectrometers and management framework are presented in brief with few specific examples of projects for industrial clients involving the characterization of materials.The existing NRU reactor is a multipurpose facility that has been used for many in-core and external experiments for almost 50 years since it began operation in 1957. Its capabilities have been vital for characterizing and testing materials for the Canadian nuclear industry. It has also supported for decades on materials research using the suite of neutron spectrometers, and is the world's largest producer of medical isotopes. Planning a replacement is a complex task affecting many areas of science and technology. Many lessons have been learned during the evaluation process that may be of interest to the international research reactor community

1.   INTRODUCTION

In Canada, for almost 50 years the NRU research reactor has been used in the characterization and testing of materials of interest to academia and industry. NRU is a multipurpose research reactor with several important features. It has a large-volume core: 200 possible fuel positions arranged in a vertical cylinder approximately 3 m high and 3 m in diameter. It has on-line fuelling capability: a fuelling machine can remove old rods and insert new ones while the reactor operates. It is cooled and moderated with heavy water. It is currently rated at a thermal power of 135 MW and has a thermal neutron flux of around 3.5×10^{14} n/cm^2/s.

The large volume of the core and the level of neutron flux allow NRU to support several scientific activities at the same time:

- in-core fast neutron irradiation of materials

- in-core testing of full-size fuel bundles

- extra-core operation of a suite of neutron spectrometers

Two government agencies manage the research facilities in the NRU reactor. Atomic Energy of Canada Limited (AECL) own and operate the reactor. AECL are responsible for the in-core research programmes, testing fuels and materials. They also oversee the in-core production of radioisotopes used in medicine and industry; NRU is the world's largest supplier of medical isotopes. The National Research Council of Canada (NRC) are responsible for the Canadian Neutron Beam Centre (CNBC) and its the suite of spectrometers. NRC manage the CNBC as an international user facility in a similar way to the *Institut Laue-Langevin (ILL)* in France or the *NIST Centre for Neutron Research* in the USA. The only facility of its kind in Canada; the CNBC is widely used by universities across Canada and around the world.

The CNBC enables three main client groups to exploit the special knowledge about materials that can be obtained only by neutron scattering experiments.

Funding for NRC-CNBC comes from the Canadian federal government, the university funding body NSERC, and fees from industry users. Fees are only charged to those users who want to keep the results of a project proprietary. Academic researchers who will publish in the open literature are not charged fees to access the facility, that cost is covered by federal funds and NSERC.

Clients and Activities	Fraction of Operation	Achievements
Canadian Universities *Academic research* *Education of highly qualified personnel*	50%	More than 20 Canadian universities coast to coast More than 50 Canadian faculties including physics, chemistry, biochemistry / biophysics, geology, materials science and engineering Between 35 and 40 graduate students access the laboratory annually, plus post docs, and professors
International collaborators *Global reach* *Strategic R&D*	25%	Canadian interaction with foreign universities, government laboratories, industries A resource for scientists from government labs A resource for scientists from NRC Institutes
Industry *Direct socio-economic impact*	15%	Fee-for-service work, on a full cost recovery basis Main industry sectors: aerospace, energy (nuclear, oil & gas), automotive, steel, aluminum Projects address important issues such as fitness-for-service, regulations, public safety, competitiveness, market penetration
NRC-CNBC *Internal development projects*	10%	Development of new neutron beam instruments, methods and applications

Neutron scattering is a versatile technique for the characterization of materials. Consequently, the NRC-CNBC and other neutron labs world-wide engage in a wide spectrum of science including physics, chemistry, biosciences, earth-sciences, materials science and engineering. Because NRU is situated at a large nuclear engineering laboratory, it has been possible to undertake some specialized neutron scattering projects on materials and components, some highly radioactive, of interest to the nuclear industry.

2. RESIDUAL STRESS MEASUREMENT

The use of neutron diffraction to address engineering questions about material properties and behaviour, started in the mid 1980s [1-3]. That activity has been encouraged and developed at NRC-CNBC over the past two decades. Industrial research projects have included many different types of experiment over the years: measuring temperature, phase composition, texture, and response to applied loads in various industrial components. The most common type of experiment required by industrial researchers however, is the non-destructive mapping of residual stresses in full-scale engineering components.

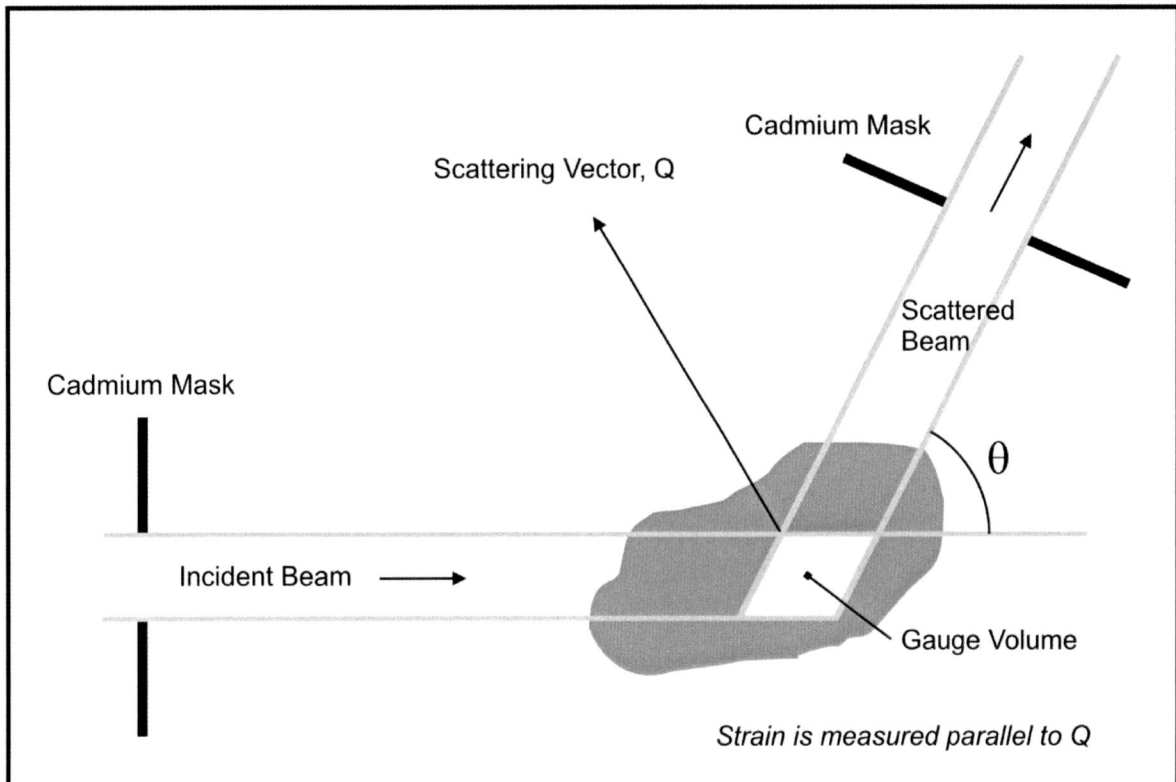

Fig. 1: The use of cadmium masks to shape the neutron beam enables strain measurements to be made on a defined "pixel" within the specimen, called the gauge volume.

Residual stresses can be generated within an industrial component by the manufacturing processes that the component goes through. Forging, stamping, welding or bending operations, and subsequent heat treatments, all of which are associated with inhomogeneous, inelastic deformation of the component, can lead to a finished component with significant residual stress. During service, that residual stress can be additive to the mechanical stress the component experiences, leading to a failure of the component under an applied load that it would normally be expected to withstand.

If a sample contains regions under residual stress, the planes of atoms in those regions will be distorted from their stress-free condition. Neutrons diffracted from the strained lattice will exit from the sample at a slightly different angle than those from an unstrained lattice . So diffraction can be used to measure the lattice strain in the grains that make up the material. In effect, the grains are tiny embedded strain gauges.

The neutron beam before and after the sample passes through rectangular slits cut in neutron-absorbing material. The volume defined by the intersection of those shaped beams is called the gauge volume (Fig. 1). Only material within this volume provides information about the strain; it can be thought of as a pixel within which the strain is sampled. In an experiment, the specimen is moved by a computer-controlled translation system to position that gauge volume at locations of interest within the specimen, gradually creating a map of the strain distribution. In broad terms that map is like a plot from a finite-element model, except that neutron diffraction can reveal the *actual strain* not a prediction.

Residual stress measurements have been made at NRC-CNBC for clients in many industry sectors: automotive, aircraft, steel production, railroad, gas storage, spacecraft and nuclear, usually for components whose failure would have significant consequences in terms of public safety or economic loss.

2.1 Residual Stress in Nuclear Pipework

CANDU nuclear reactors are pressurized heavy water reactors in which the fuel is held in hundreds of individual channels. The heavy water coolant is delivered to those channels via pipes, from 1.5" to 3.5" in diameter, called feeders. A leak from such a pipe will lead to a temporary shutdown of the reactor while the problem is fixed. Such a shutdown can result in millions of dollars of lost electricity production.

When a feeder developed a crack at a CANDU station, a series of experiments were conducted to understand the factors that had led to the crack occurring. Measurements of residual stress in a typical feeder bend were made first by neutron diffraction because the non-destructive technique allowed further types of characterization experiment to take place on the same feeder afterwards.

Initially measurements were made around the circumference of the feeder near the inner surface, those showed a clear peak in the hoop stress at a certain location. Readings were then made at locations through the wall thickness at that position. The through-wall measurements showed a steep gradient in the hoop component of stress, with tensile stresses that approached yield near the inner surface. Examination of the cracked feeder bend along with these measurements, indicated that residual stress had played a major role in the failure.

Subsequent to that investigation, detailed studies have focused on a selection of feeders manufactured by different methods [4]. It was determined that although the various pipe bends meet specifications in terms of material, bend angle, cross-sectional shape, and wall thickness, the resulting stress distributions are markedly different.

3. CHARACTERIZATION OF RADIOACTIVE COMPONENTS

It is common for neutron scattering experiments to take advantage of the penetrating nature of neutrons and place the sample that is being examined within a container that creates desirable conditions of pressure, temperature, load or other environmental parameters. A slight twist to that scenario is a piece of equipment developed at NRC-CNBC enabling experiments on highly radioactive samples. Called a *shielded cell*, it is designed to protect staff from the radiation emitted by radioactive samples, rather than to contain an artificial environment around the sample.

The shielded cell is essentially a lead box containing a translator on which a sample can be mounted. The sample can then be moved in two axes inside the cell to allow measurements to be made at precise locations. A number of ports are built into the cell walls with plugs that can be removed to allow the incident and scattered neutron beam to pass through. Radiation from the sample that shines out of the cell when the ports are opened up is only directed into the neutron detector and its shielding, or back up into the spectrometer's monochromator cavity inside its shielding drum, representing no radiological hazard.

The shielded cell was tested with a 400 GBq Co-60 source and found to give acceptable radiation fields in the experiment area. In a recent project, samples of welds in low cobalt stainless steel from a reactor structure were measured in the unirradiated state, and again after irradiation with fast neutrons.

Using the shielded cell NRC-CNBC has been very successful with industrial research in the nuclear industry. Studies of structural components from reactors have compared material before and after irradiation to determine if material properties degrade over time beyond acceptable limits. Such studies obviously have significant economic impact, affecting decisions of the operating lifetime of nuclear electrical generating stations. As in other examples an important aspect of the neutron scattering results is the fact that they are not models or simulations, but an actual measurement of material conditions. In a highly regulated environment such as the nuclear industry, such definitive data adds a high level of confidence in the resulting decisions by management or regulatory organizations.

4. NUCLEAR FUEL RESEARCH

NRC-CNBC has a powerful powder diffractometer, which is widely used by the crystallography, chemistry and geoscience communities. Neutron powder diffraction has proved a valuable tool for the development and testing of nuclear fuels. Powder diffraction has become a routine step in the qualification of research reactor fuel at Chalk River. Phase analysis of various uranium alloys and ceramics enables scientists to track the behaviour of fuel under conditions of temperature or radiation. Fresh fuel can be analyzed at intermediate points of the production process, to determine what effect a manufacturing step has on the structural phase of the uranium-bearing material. A second shielded cell enables powder diffraction experiments on used and partially-used fuel.

A recent powder diffraction experiment involving scientists from NRC and Atomic Energy Canada Limited (AECL) focused on Uranium-Molybdenum-Aluminium fuel. The sample was irradiated fuel (20% burn-up). Because NRU is a multipurpose neutron source, the test fuel was irradiated in the core and then examined on the powder diffractometer, all in the same research facility.

The aim of the experiment was to help understand the swelling behaviour of the fuel with respect to the crystalline and amorphous phases that form within the fuel during irradiation. There was some expectation that the irradiated fuel would be fully amorphous following irradiation. However the samples were found to contain a number of crystalline phases [5] that were identified in the data analysis using the GSAS Rietveld refinement software.

5. RESEARCH ON MATERIALS FOR HYDROGEN STORAGE

Because of the particular sensitivity of neutrons toward hydrogen and its isotopes, there are areas in which neutron scattering can make important contributions in the development of new hydrogen technologies related to fuel cells and their use. One important challenge in making a viable hydrogen-powered vehicle, is how the hydrogen is stored. A number of materials have been suggested as candidates for storing hydrogen in a solid state, and release gaseous hydrogen on demand. The advantages of such systems are in safety and size. A solid hydrogen-bearing compound may not have the same potential for fire hazard as pressurized hydrogen gas. Storing hydrogen as part of a solid compound would also be an efficient, compact solution, compared to gaseous storage.

At NRC-CNBC we have developed a capability to do research on materials that are candidates for hydrogen storage applications. To aid this research, a dedicated piece of apparatus that can hold potential hydrogen storage materials in the neutron beam while hydrogen (or deuterium) is cycled in and out of them has been developed. It will allow scientists to study the performance of these materials, and determine where hydrogen is taken up within the atomic structure of the storage medium. This is a good example of neutron diffraction on a sample in actual working conditions. The project also illustrates the close connection between energy and materials as research topics. The two are often closely connected, energy technologies relying heavily on specialized materials for their success. Gas combustion, electricity distribution, nuclear fission and fusion technology or fuel cells can all benefit from advances in materials

6. INSTRUMENT DEVELOPMENT

NRC-CNBC sees an essential part of its mandate is to develop and improve the technical capabilities of the neutron scattering laboratory which NRC manages as a national science facility.

A neutron reflectometer is currently under construction at NRU. Reflectometry is a technique that can probe thin films and surfaces, or buried interfaces in layered materials. Using reflectometry, scientists can measure larger structures (up to 200 nm) than is possible with diffraction, which typically measures structures of around 0.2 nm. Neutron reflectometry has been growing in popularity in Canada, under the stewardship of NRC-CNBC. Experiments for the past 10 years have been carried out by adapting a triple axis spectrometer to operate in reflectometry mode. The new instrument that will be completed later this year will enable greater scientific activity in this emerging field. The reflectometer project is being led by the University of Western Ontario, with 12 other university partners, who secured funding from the Canada Foundation for Innovation, NRC and the province of Ontario, demonstrating the demand for the new instrument across the Canadian academic community.

A *white microbeam* scattering capability has been developed at NRC-CNBC in partnership with Oak Ridge National Laboratory in the USA. It comprises equipment that attaches to the engineering diffractometer, producing a narrow (1 mm x 3 mm), polychromatic neutron beam from the reactor. That beam is then focused using a pair of Kirkpatrick-Baez super-mirrors to a cross-section less than 0.1 mm x 0.1 mm. Laue scattering of the beam is detected on a specially-designed 2-D neutron image plate. The result appears as an array of spots corresponding to the Bragg reflections of specific wavelengths within the neutron beam spectrum that scatter fromcrystal planes satisfying the Bragg condition. This very narrow beam can diffract from a single crystal within a polycrystalline sample such as a metal: a useful tool for investigating stress or strain at a microscopic level.

7. A FUTURE RESEARCH REACTOR IN CANADA

The largest issue for Canada in the field of neutron scattering, concerns the research reactor itself. As the stewards of the neutron scattering facilities at the NRU reactor, the National Research Council (NRC) recognizes that NRU is now approaching the end of its life, having operated since 1957. NRC is Canada's federal science agency, with institutes that focus on a wide range of scientific issues including biotechnology, aerospace, information technology, astronomy, environmental technology, microstructural sciences and manufacturing technology. NRC has been studying the issues related to research reactors and their place in the national science infrastructure.

It is important to understand the way research reactors have developed in Canada to give some perspective on what demands may be made of a new facility. NRU was designed and built in the 1950s when Chalk River was a thriving centre for the physical sciences. NRU was built as the most powerful research reactor at that time: 200 MW thermal power and an intense thermal neutron flux. It was also designed to be very flexible, with many beam tubes for experimental equipment and a versatile core design to accommodate in-core test facilities. Over its life it has operated with high-enriched, natural and low enriched fuel.

7.1 A multipurpose science facility

Unlike many countries that built separate research reactors for nuclear technology development and neutron scattering, Canadian reactors, NRX then later NRU, were designed to be flexible enough to support both those activities and more. The NRU reactor has proved essential for characterizing and testing materials for the Canadian nuclear industry. It has also supported decades of materials research using the suite of neutron spectrometers, following

Brockhouse's pioneering work in the field. NRU is also the world's largest producer of medical isotopes.

Those three activities alone are quite diverse, affecting many sectors of society. Over its lifetime, though, NRU has supported other activities as the demand arose: fundamental physics experiments and neutron radiography for example. Because of its multipurpose nature, Canada's research reactor is more than a 'nuclear facility', it is a 'science facility' serving national needs in health, science, industry, education, energy, environment and trade.

Canada has been well served by the NRU reactor and a replacement facility will need to support the same breadth of science and technology; activities within and beyond the traditional nuclear field.

7.2   Three replacement concepts

There are three ways in which the capabilities of NRU can be continued in Canada into the 21st century: a replacement multipurpose facility can be built, separate single-purpose facilities can be built, or NRU itself can be extensively refurbished. These three concepts were the focus of a cost-estimate study that NRC commissioned from AECL in 2004.

That study identified some interesting facts. The capital cost of a refurbishment that would give comparable lifetime to a new facility was estimated at around \$430M whereas a new construction was approximately \$650M. Constructing separate facilities for nuclear R&D, neutron scattering and isotope production was unsurprisingly the highest capital cost, due to triplication of structures and systems. The principle conclusion was that design and construction of the civil structures and the large mechanical systems dominated the capital cost. The footprint of the reactor core only changes the cost by an incremental amount. To construct a reactor hall with cooling and control systems for a compact core (less than 1 m^3) for neutron scattering, or for a large multipurpose core (20 m^3 as in NRU) entails many similar costs. One concept is not double or triple the cost of the other.

8.   CONCLUSION: A National Mission

Canada has the smallest population amongst the G8 countries, and with the corresponding smaller tax-base it needs to apply some ingenuity in its choices for major science facilities. The NRU reactor reflects such ingenuity, and has established itself as Canada's most productive science facility as a result: supporting a \$5 billion nuclear industry, enabling world-class neutron scattering research and producing isotopes for more than 20 million patients annually. Canada's next research reactor will be a significant national investment, and one that demands a similar level of productivity.

REFERENCES

[1]   HOLDEN, T.M.,et al., Proc. 5th Canadian Conf. on NDT (1984).
[2]   ALLEN, A.J., HUTCHINGS, M.T., WINDSOR, C.G.,ANDREANI, C., Adv. In Physics **34** (1985) 445.
[3]   HOLDEN, T.M., Proc. Intnl. Symposium on Neutron Scattering (1987).
[4]   YETISIR, M., ROGGE, R.B., DONABERGER, R.L., Proceedings of ASME Pressure Vessel and Piping Division Conference, PVP2005-71581 (2005).
[5]   CONLON, K., SEARS, D., 10th Topical Meeting on Research Reactor Fuel Management: European Nuclear Society (2006).

Design features and current status of HTR-10GT

WANG JIE[*], HUANG ZHIYONG, YANG XIAOYONG, SHI LEI, YU SUYUAN

Institute of Nuclear and New Energy Technology, Tsinghua University,
Beijing, China

Abstract: The 10MW high temperature gas-cooled reactor (HTR-10), a pebble bed type reactor, is the first reactor worldwide which has inherent safety features. It reached its first criticality in 2000 and began to operate on full power in 2003. The power conversion system of the HTR-10 is a steam turbine generator system. Based on the success of the HTR-10, a new project, the gas turbine power conversion system coupled with the HTR-10, was launched, which is denoted as HTR-10GT. The HTGR gas turbine cycle is expected to have higher efficiency and better performance theoretically. Therefore, for the HTR-10GT, the gas turbine direct cycle is utilized instead of previous steam generator and steam turbine. The arrangement of helium turbine and electric generator is selected as single shaft configuration with a gear box between the two machines, for which the turbine speed is designed as 15000r/min and the generator is 3000 r/min. The rotors are supported by active magnetic bearings to avoid the contamination from any lubricant. The reactor core outlet temperature is designed as 750 □ and inlet 330 □. The plant power is controlled by adjusting reactor control rods and helium density simultaneously, so that it can remain high efficiency even in partial loading. The overspeed of the turbomachine is restricted by opening bypass valves to reduce pressure difference between the inlet and outlet of turbine. This presentation will show the main design features and current status of the HTR-10GT project.

1. TECHNICAL BACKGROUND

The High Temperature Gas-cooled Reactor (HTGR) is a representative of the next generation of nuclear system. In China, the 10MW High Temperature Gas-cooled Test Reactor (HTR-10) was build in the Institute of Nuclear and new Energy Technology (INET), Tsinghua University, which reached its first criticality in 2000 and begun its full power operation in 2003(Wu, et al., 2002). Currently, the long term operation and the safety demonstration tests of the reactor are carried out in order to gain the operation experience and verify the inherent safety features of the HTGR. Based on the success of the HTR-10, two projects have been recently launched to further develop the HTGR technology. One is a prototype modular plant denoted by HTR-PM to demonstrate the commercial capability of the HTGR power plant, of which the reactor core power is 380 MW and the power conversion system is steam turbine generator (Zhang, et al., 2004). The other is a gas turbine generator system coupled with HTR-10 denoted by HTR-10GT to demonstrate the feasibility of the HTGR gas turbine technology (Wang, et al., 2004).

For the HTR-10GT project, the gas turbine generator is expected to be installed inside the HTR-10 primary system, instead of previous steam generator and steam turbine, to simplify the reactor system and elevate the plant efficiency. The initial basic design was a joint one by the Institute of Nuclear and new Energy Technology (INET), Tsinghua University, and State Unitary Enterprise I.I. Afrikantov, Experimental Design Bureau of Mechanical Engineering (OKBM), Russia. After that, the engineering design, component R&D and key technology research are carried out by the INET. This paper provides the main design features and current status of the HTR-10GT project.

2. GENERAL DESIGN

Figure. 1 shows the layout of the HTR-10 primary system. The left side pressure vessel contains the reactor core and the right side one contains the steam generator and helium circulator, which are connected by a horizontal hot gas duct pressure vessel. The nuclear energy generated in the core is transported to the steam generator by the helium circulation

---

[*] Corresponding author. Tel: +86-10-6279-4678; Fax: +86-10-6277-1150.
E-mail address: wjinet@tsinghua.edu.cn

within the primary system forced by the helium circulator, which forms the nuclear steam supply system.

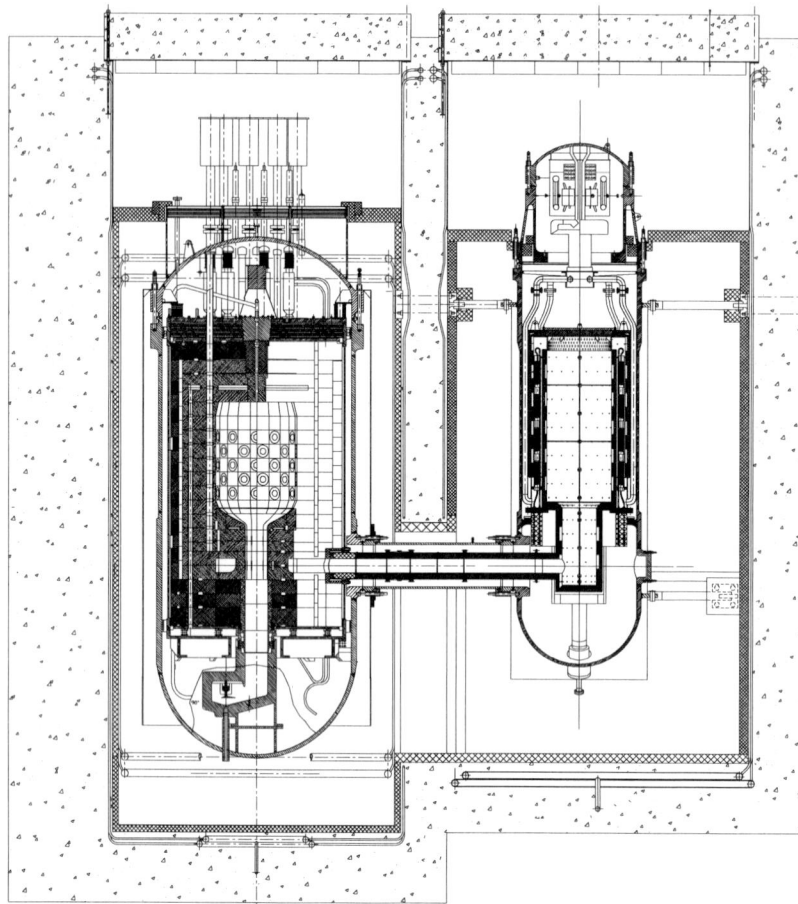

Fig. 1. Layout of the HTR-10 primary system

For the new project of the HTR-10GT, the previous pressure vessel of steam generator in right side will be removed and a new pressure vessel containing helium turbine and generator will be installed, shown in Fig. 2. The new pressure vessel is called as power conversion vessel (PCV) to convert reactor core energy to electric energy finally. The turbomachine is designed in the vertical axis. From lower to up, these machines are turbocompressor, gear box and electric generator, for which the speed of turbocompressor is designed as 15000 r/min and the generator is 3000 r/min. The turbocompressor and generator are supported by the active magnetic bearing system to prevent the primary system from lubricant contamination. The PCV is divided into two chambers, of which the up one is generator chamber keeping 65 □ required by generator and the lower one is inside the primary system. The various heat exchangers are installed around the turbocompressor.

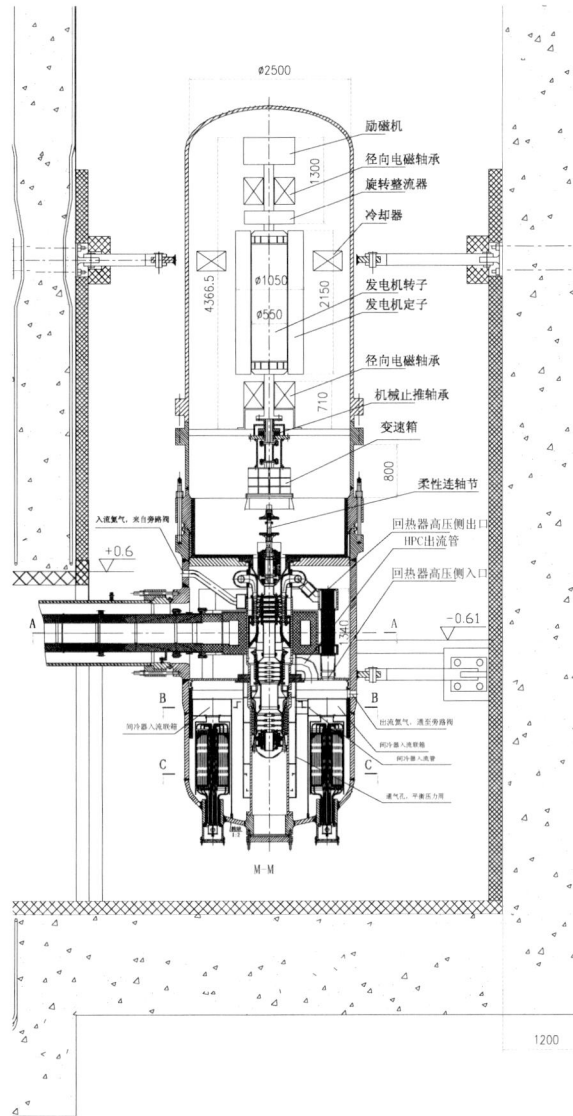

Fig. 2 Layout of Power Conversion Vessel

Helium flow chart is depicted in Fig. 3. The helium from the core with high temperature and pressure expands in the turbine and drives turbine with the compressors and generator together. The exhaust helium enters the recuperator to heat helium in the other side from the compressor outlet. Then the helium is further cooled in the precooler and pressurized to high pressure through the two-stage compression process with intercooling. The high pressure helium is preheated through the other side of the recuperator as mentioned above and then enters into reactor core to be heated to complete the Brayton cycle. The main parameters of the HTR-10GT in the mode of operation at full power are shown in Table 1.

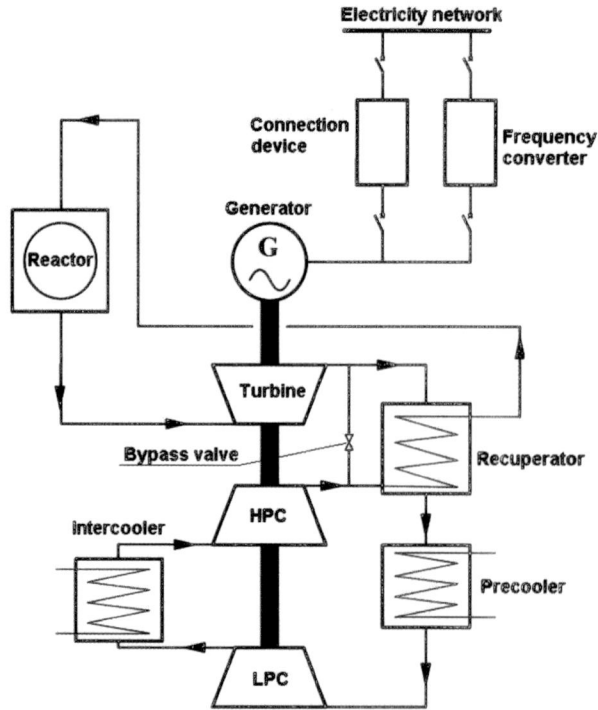

Fig. 3 Thermodynamic cycle diagram of HTR-10GT

TABLE 1. MAIN PARAMETERS OF HTR-10GT

Reactor Core	
Power, MW	10
Inlet /outlet Temperature, °C	330 / 752
Inlet / outlet Pressure, MPa	1.53 / 1.52
Mass flow rate, kg/s	4.55
Turbomachine	
Turbocompressor speed, r/min	15000
Turbine expansion ratio	2.2
Compressor ratio for HPC	1.58
Compressor ratio for LPC	1.58
Generator speed, r/min	3000
Recuperator	
Power, MW	15000
Helium temperature at inlet/outlet（LP）, °C	494/278
Helium pressure at inlet/outlet（LP）, MPa	0.687/0.682
Helium temperature at inlet/outlet（HP）, °C	109/330
Helium pressure at inlet/outlet（HP）, MPa	1.605/1.604
Precooler	
Power，MW	5.94
Helium temperature at inlet/outlet，°C	278/35
Helium pressure at inlet/outlet，MPa	0.682/0.676
Intercooler	
Power，MW	1.79
Helium temperature at inlet/outlet，°C	108/35
Helium pressure at inlet/outlet，MPa	1.040/1.031

HTR-10GT is based on the closed Brayton thermodynamic cycle; there are the following distinguishing features to be mentioned:

1) It demonstrates the feasibility of direct Brayton cycle with nuclear reactor as heater, helium as the working fluid.

2) Helium moves around in a closed loop, which implies that no helium is consumed in the power generation process, it merely acts as an energy carrier.

3) The helium closed cycle gas turbine gains the benefits on low Reynolds number, low Mach number and not serious stall problem.

4) Because of higher heat transfer, it is possible to design helium compact heat exchangers.

5) It makes use of a recuperator to recover heat that would otherwise have been rejected to cooling water. The recovered heat is transferred elsewhere in the system thereby reducing the heat required from the heat source and ultimately increasing the thermal efficiency of the plant.

For a closed thermodynamic cycle, entropy is the quantitative measure of the amount of thermal energy not available to do work, changes of entropy must therefore always be positive, yet a negative change is indicated during heat rejection. Turbine and compressors work with expansion and compression at nearly constant entropy, and heat addition and rejection at almost constant pressure. If the process is adiabatic (no heat transfer), but not reversible (with losses), the entropy increases. Fig.4 shows the relationship between temperature and entropy, and Fig.5 shows the relationship between pressure and specific volume of the real HTR-10GT thermodynamic cycle.

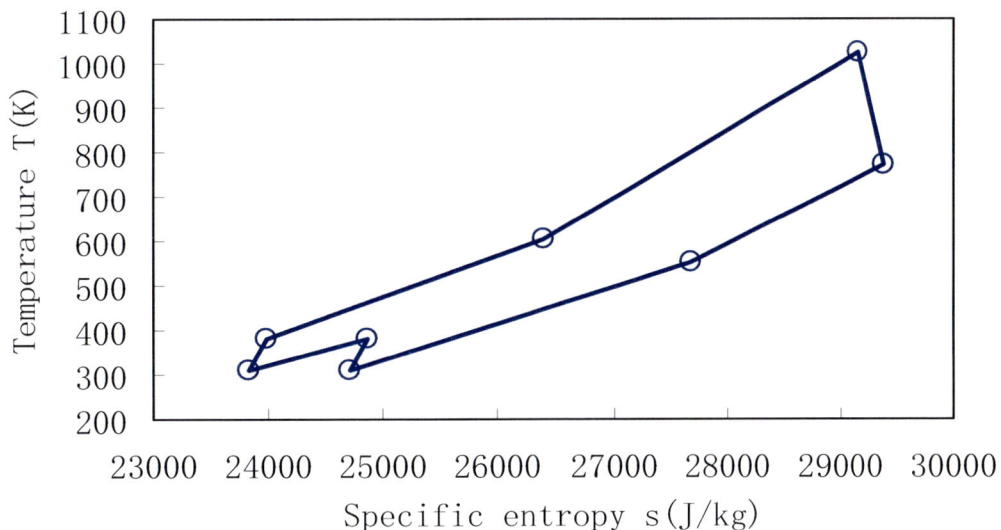

Fig.4 Relationship between temperature and entropy

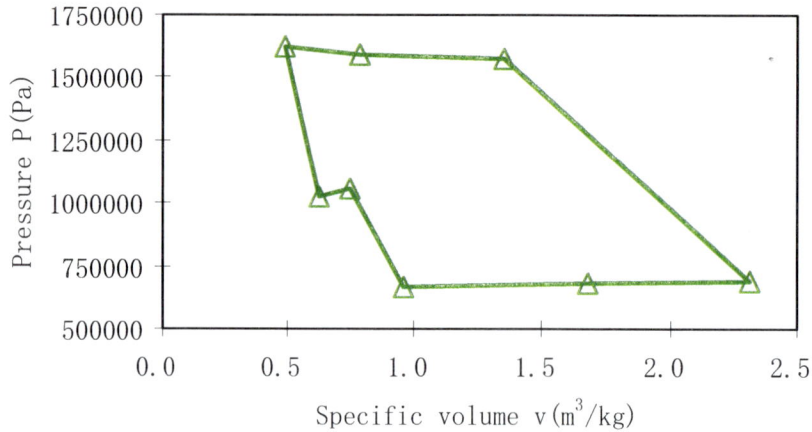

Fig.5 Relationship between pressure and specific volume

Figure 6 shows the relationship between the relative reactor power and the relative generator power during the start-up to full power stage. At the beginning, the reactor is set in sub-critical state. All the plant systems and equipment are checked and prepared for start-up. During the reactor start-up, the helium mass is necessary to ensure primary circuit filling pressure similar to atmospheric. The generator operates in motoring mode, powered by the frequency converter, and the rotor speed increases to 15000 r/min. Then the reactor reaches its criticality and increases its power smoothly. Helium supply to the turbine is started, turbine power increases with the increase of helium parameters. When the turbine power reaches the level sufficient to ensure operation of compressors, generator is switched over to the mode of electric power generation and provides electric power to the external grid. By further increasing reactor power up to 36% of nominal and by adding helium mass into the circuit, the power is increased to the minimum level of the automatic electric power control range 30%. The power conversion system can operate at any steady-state level of power within the automatic adjustment range from 30~100% of the nominal power.

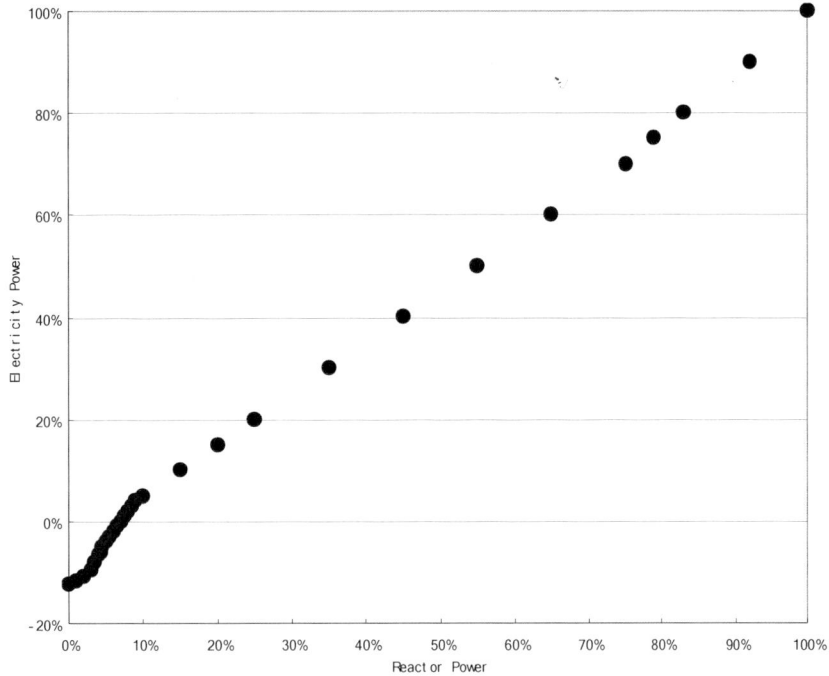

Fig. 6 Relationship between generator power and reactor power

3. COMPONENTS' R&D

Inside the Power Conversion Vessel, the main components can be divided into two groups. One is rotating group containing turbocompressor, gear box and generator, for which the turbocompressor and generator are supported by the active magnetic bearing system. The other is stationary group containing heat exchangers, pressure vessel, metal works, valves, penetrations, etc. The components' R&D is in process now. Among these components, the turbocompressor is most important one which meets many technical difficulties. Hence we firstly focus our attention on the turbocompressor.

The turbocompressor in HTR-10GT is to convert nuclear energy into electric energy in direct gas-turbine cycle and provide primary coolant (helium) circulation. The turbocompressor contains three components, turbine, low pressure compressor and high pressure compressor. The compressors and turbine are vertically arranged in the lower cavity of power conversion vessel as shown in Fig. 7. From bottom to top, there are lower active magnetic bearing and catcher bearing, low pressure compressor, high pressure compressor, turbine, buffer seal, repair seal, upper active magnetic bearings and catcher bearings. At the top of turbocompressor shaft, a flexible coupling is used to connect the reduction gearbox and transfer the torque between the turbocompressor and generator.

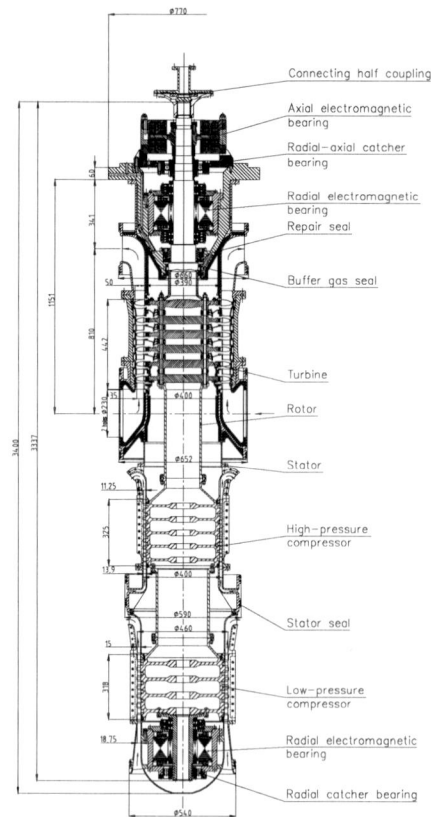

Fig 7 The arrangement of turbocompressor

The features of helium turbocompressor are lower compression ratio and expansion ratio, more stage numbers, higher rotating speed, larger hub diameter and shorter blade height than air turbocompressor. The thermophysical properties of helium greatly influence the aerodynamic parameters of turbocompressor, such as specific works, stage numbers and efficiency. Since the specific heat of helium (about 5200 J Kg-1 K-1) is much larger than that of air (about 1100 J Kg-1 K-1), the specific works for helium are larger than that for air. And helium is difficult to be compressed or expanded; the stage pressure ratio or expansion ratio of helium turbocompressor is quite low, even its load coefficient is higher than that of air. It leads to the stage numbers of helium turbocompressors much more than that of air. In HTR-10GT, the pressure ratio of low pressure compressor and high pressure compressor is about 1.58; the whole pressure ratio of helium compressor is 2.42 and the expansion ration of turbine is 2.20. On the other hand, the sound velocity of helium is about two times larger than that of air, the maximum Mach in turbocompressor is about less than 0.5, so the aerodynamic loss is correspondingly low and the efficiency of helium turbocompressor is quite high. The adiabatic efficiencies of low pressure compressor, high pressure compressor and turbine are 84.5%, 84.5% and 86.5% respectively. The single shaft arrangement can prevent the over speed of turbocompressor when the generator lose its load suddenly.

Another important feature in design of helium turbocompressor is its short blade height. The blade height is determined by the volume flow rate and the helium flow rate in turbocompressor is almost directly proportional to the reactor power; in HTR-10GT it is about 4.55 kg/s. So it must be careful to choose the appropriate rotate speed in order to increase the blade height and avoid the difficult in design and fabrication. The speed of turbocompressor at nominal operation point is designed as 15000 r min and its maximum speed is 18000 r /min in HTR-10GT. On the other hand, the less are stage numbers, the smaller is the axial size of turbocompressor. So the large hub diameters and high stage load coefficients are chosen to promote the peripheral velocity to reduce the axial height of turbocompressor. The fundamental parameters of turbocompressor are summarized in Table2.

TABLE 2. The Fundamental Parameters Of Turbocompressor

Component	LPC	HPC	Turbine
Power / MW	1.74	1.77	5.73
Flowrate / kg/s	4.76	4.77	4.66
Inlet temp. / ℃	23.9	26.7	750
Outlet temp. / ℃	94.3	97.8	502
Inlet press. / MPa	0.65	1.00	1.50
Outlet press. / MPa	1.03	1.58	0.68
Efficiency / %	84.5	84.5	86.5

The key aspects in design of turbocompressor include the structural design and aerodynamic analysis. Different design schemes are compared by the aerodynamic analysis to get optimal stage numbers, stage load coefficients, blade profiles, efficiency and so on. The aerodynamic analysis optimizes the three-dimensional flow field inside turbine by Computational Fluid Dynamics (CFD) programmes in partial and full load; furthermore it also provides the aerodynamic performance of turbocompressor. Structural design of rotor, stator, casing and other components guarantees the component strength, reliability, lifetime and vibration analysis. As a result of design and analysis, the geometrical characteristics, mass and size characteristics of the turbocompressor, are obtained as shown in Table 3 and Table 4 respectively. The total height of turbocompressor shaft is about 3400mm, and the maximum diameter of the assembly is about 770mm.

TABLE 3. THE GEOMETRICAL CHARACTERISTICS OF THE TURBOCOMPRESSOR

Component	LPC	HPC	Turbine
Blade height rotating / stationary blades / mm	18.8 /15.0	13.9 /11.3	35 /50
Tip diameter / mm	460	400	490
Net length / mm	318	325	442
No. of Stages	6	8	6
No. of rotating / stationary blades	650 /733	1014 / 1122	372 /252

TABLE 4. THE MASS AND SIZE CHARACTERISTICS OF THE TURBOCOMPRESSOR

Rotor weight / kg	540
Rotor length / mm	3337
Total weight / kg	2750
Total length / mm	3400
Max diameter / mm	770

The most difficulties in design of turbine are how to calculate the temperature field in turbine and design the coolant system. It directly influences the operational reliability of turbine. The thermal stress caused by high temperature is so prominent that it should be taken into account during the strength analysis and lifetime analysis.

One of the aims of HTR-10GT is to verify the aerodynamic design of helium turbocompressor. A closed areodynamic test facility with air or helium was set up to verify the design of compressors. There are two important reasons for such test. Firstly the thermophysical properties of helium are very different from that of air, and the aerodynamic design of turbocompressors is based on the experience gained in air. Whether these experiences can be applied in helium circumstance and how about the accuracy of aerodynamic design? On the other hand, the computational performance of turbocompressor should be checked by the experimental results. The performance test of turbine and compressor is usually conducted in air because the helium is expensive and easy to leak out. So the test data in air must be converted to the performance data in helium by similarity criterion. Whether can the similarity law be applied in different gas such as helium and air? How about the conversion of performance data between air and helium? Such questions must be answered before the engineering design and manufacture of turbocompressor.

The test facility is a closed loop consisting of air and helium gas storage tanks, desiccator, flow regulating valve, membrane compressor, gas heater and cooling system, pressure regulation system, test section, reduction gearbox, motor and control system as shown in Fig. 8. The helium or air is pressurized by membrane compressor, and the test blades or compressor is placed in test sector. The blades or compressor is driven by motor through the gear box to reach the design rotating speed, 15000 r /min. The gear box is lubricated by oil and the dry gas seal is used to prevent the shaft gap from the leakage of helium.

Fig 8 The picture of testing loop for helium compressor

Several experiments are and will be carried out on it. First is to test the design blade in helium and air, the second is to test a stage of blades in high pressure compressor in helium and air; and the other one is to test the whole high pressure compressor in air. The aims are to get the helium and air behavior for helium blade design, to get the performance of compressor, to check the error of similarity law in data conversion. Now the fabrication and installation of testing loop are completed, and the blade experiments in helium and air are currently in process.

Additional requirements for generator and reduction gearbox in rotating component group during the start-up and shut-down processes are special. At the beginning stage of start-up (or shut-down) process, the generator is used as motor to drive the turbocompressor, and then it works as the generator in load operation. Design scheme of synchronous generator is adopted with two working mode, motor and generator. A frequency-converter is used to start generator in motor mode. Corresponding to the operating mode of generator, reduction gearbox also has two working mode, reduction gearbox and speed-up gearbox. Therefore both sides of the tooth are designed as working surface. Reduction gearbox connects with the generator and turbocompressor by membrane couplings.

The design of modular heat exchangers, including recuperator, intercooler, precooler, are approaching to end now. Penetrations, metal works and pressures vessel are all in their design phase. Almost all components of HTR-10GT will be developed and fabricated by Chinese research institutes and companies.

4. AMB TECHNOLOGY

The active magnetic bearing (ABM) system is the key technology for the HTR-10GT project. The INET has made many efforts on the issue. Here we present the experimental research on the AMB technology for the HTR-10GT project.

4.1 Experimental programme

In order to explore the AMB engineering design and validate the technology, we have elaborately made a full experimental plan: first, a small test rig is established to test the control method for flexible rotor and accumulate experience for passing through critical speeds. Then a large size of magnetic bearing with a rigid rotor is constructed to verify the large magnetic bearing design and check its characteristics in long time operation. After the above two experiments are successful, a full scale engineering test (1:1) will be performed outside the reactor to validate all the designed properties of the AMB system. Finally, the actual turbomachine rotor system along with the AMB system will be mounted in the PCU vessel of the HTR-10 reactor.

At present, we have already built the former three test rigs and performed a serial of experiments. On the small flexible rotor test rig, we have succeeded passing through the second bending critical speed (BCS). And on the rigid test rig, we have finished the property experiment study and checked the AMB system in a continuous running more than 72 hours. The full scale engineering test for the generator rotor has been installed, and now the suspending and low speed running is achieved. Next we will continue the experiment according to our plan.

4.2 Small flexible rotor test rig

The emphasis of this experiment is to study the control arithmetic of how to pass through the bending critical speed (BCS) and to provide enough experience for the future turbomachine rotor control. The first and second BCS are designed as 300Hz and 700Hz respectively, which are higher than that of the actual turbomachine rotor. It is deliberate to increase some difficult for the control research considering the difference between small rotor in experiment and large rotor in the PCU. The structure and main parameters of the setup are shown in Fig.9 and in Table 5 respectively.

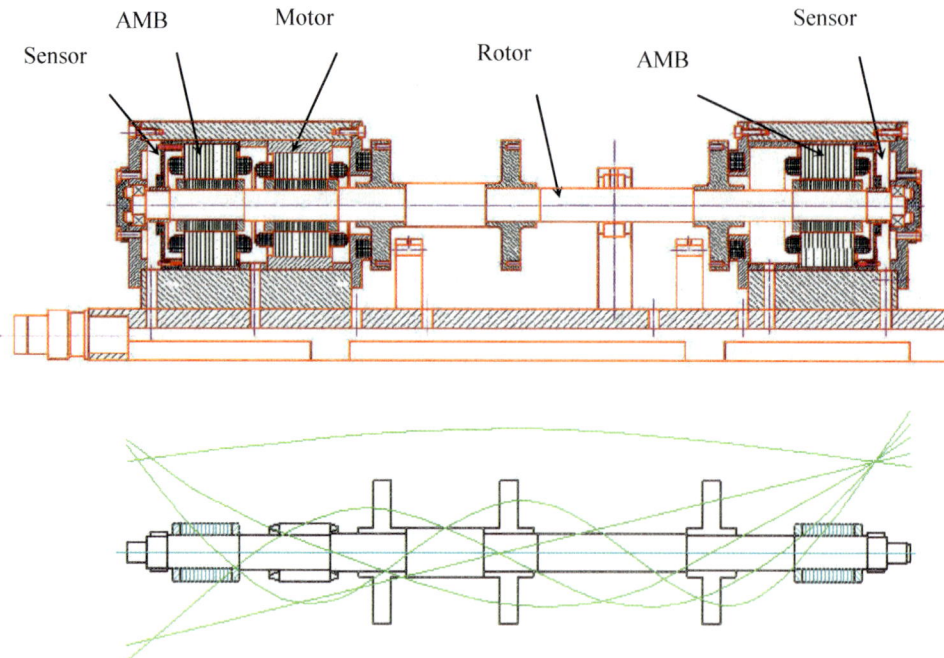

Fig.9. The structure of the small flexible test rig

TABLE 5. MAIN PARAMETERS OF THE SMALL SETUP

Rotor Mass	6.128kg
Rotor Length	613mm
Radial Moment of Inertia	0.148kg m^2
Polar Moment of Inertia	0.00379kg m^2
Air Gap	0.4mm
Coils	300n
Pole Area	320mm^2
Inductance	45.2mH

A PC controller is utilized to control the small experiment system on the famous free-charge real-time Linux system, and the sampling ratio is 10k Hz. Although the control system hardware is different from the one that will be used in the actual application, the control model and arithmetic are similar from the view point of mathematical and control. Firstly a kind of LQG controls method with phase compensating, and then a Hinf robust control was used to design the controller based on state-space modern optimal control theory.

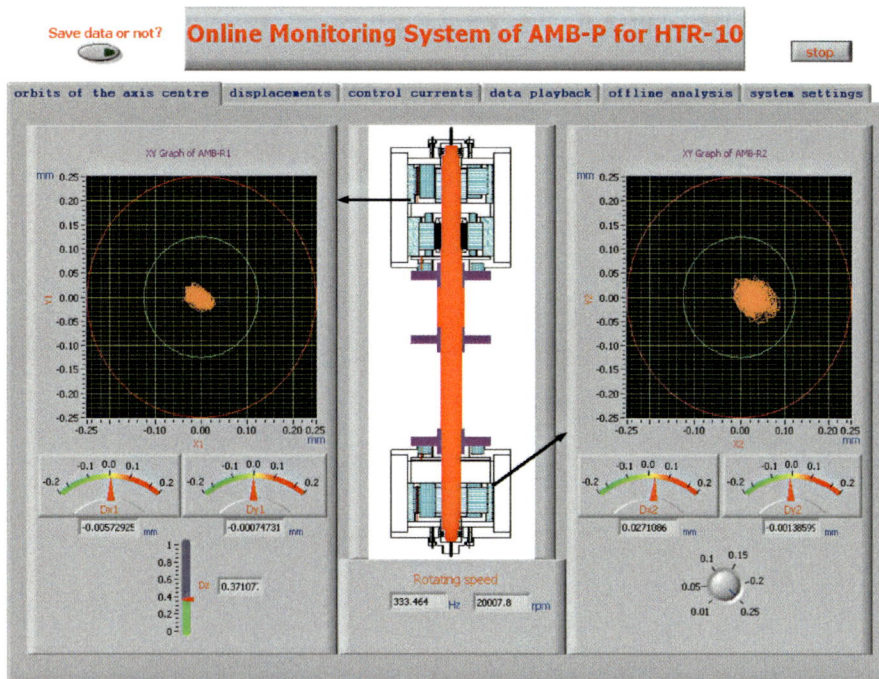

Fig.10 Main screen of the measuring and monitoring rotor system

An online measuring and monitoring system based on the VI (Virtual Instruments) technology and LabVIEW (National Instruments) platform is built for this experiment in order to detect how the AMB system works in operation and make diagnosis whether the system behaves normally or not. The main screen of this system is shown in Fig.10. User can easily and clearly know the operation status including the orbit of the axis centre, the four radial displacement signals and their spectra during passing the critical bending speed.

On the small test rig, the passing through BCS experiment was carried out elaborately. The control system around the first two bending critical frequencies of 300Hz and 700Hz has perfect control performance. The rotor passed through the second BCS safely and smoothly. Even so, the test rig can rotate at the BCS for a long time without any abnormal phenomenon. Figure 11 shows the axis loci and its frequency domain diagram in passing through the second BCS.

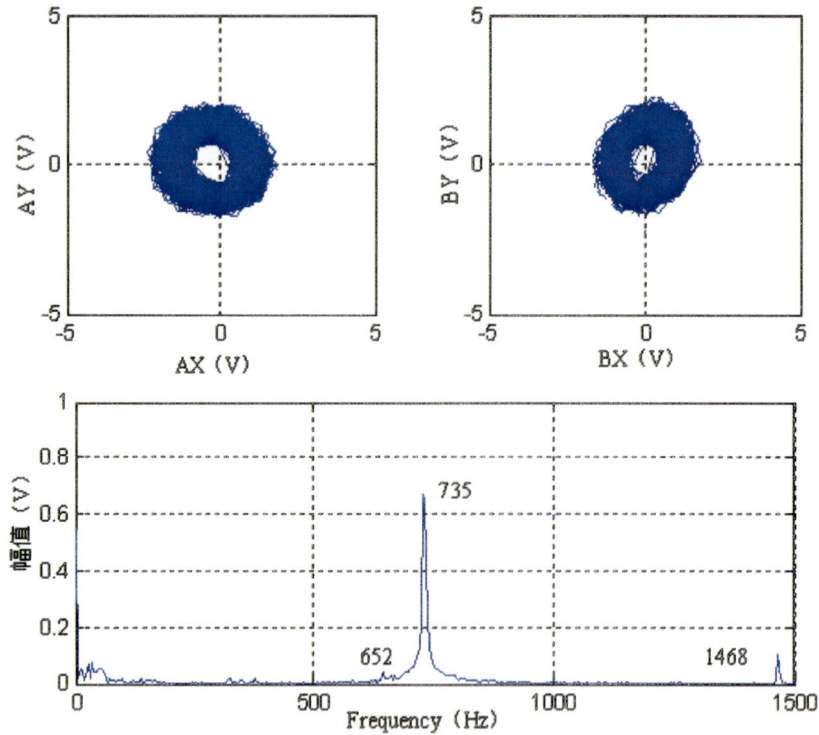

Fig.11 Time and frequency domain diagram when passing through the Second BCS

The successful passing through the second BCS verifies that the modeling and control design method is feasible and effective. We also have obtained a lot of experiences in this experiment, which will be certainly useful in the future actual tuning process for the HTR-10GT AMB system.

4.3   Large size rigid rotor test rig

This experiment aims to study the actual large size AMB's characteristics and verify the design process because of our lacking of large AMB design experience. We also want to test the performance the prototype of the power amplifiers and improve them in the next design phase. In order to simplify the other difficulties, a rigid rotor is designed to make our attention to the magnetic bearing itself. The structure layout of this test rig is shown in Fig.12.

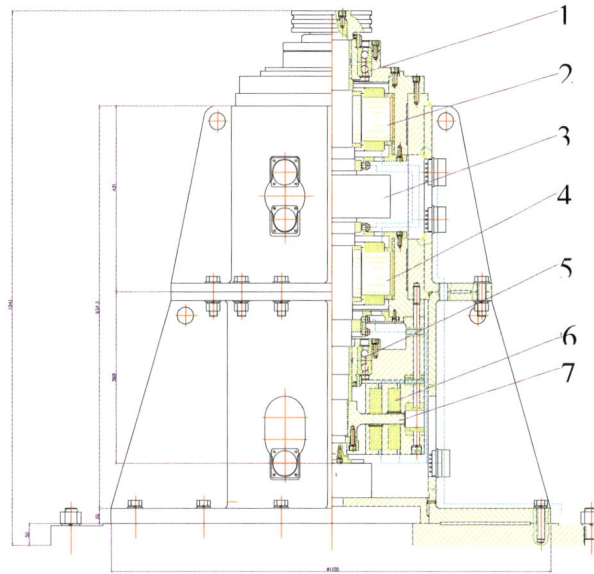

1 – upper catcher bearing; 2 – upper radial AMB; 3 – power winding;

4 – lower radial AMB; 5 – lower catcher bearing; 6 – axial AMB; 7 – thrust disk

Fig.12 System structure layout of the full size magnetic bearing test rig

A standard industrial control computer with a Pentium 4 CPU of 2GHz is selected as the main controller. The power amplifier is a kind of two H-bridge connecting in serial type with a switching frequency of 60k Hz and total power of 45k VA. The main parameters are shown in Table 6.

TABLE 6.MAIN STRUCTURE PARAMETERS OF THE RIGID ROTOR TEST RIG

Parameter	Value
Height / weight of the rotor, mm / kg	1,200 / 150
Radial AMB Lifting capacity, N	3,000
Radial gap between bearing and rotor, mm	0.7
Radial gap between catcher bearing and rotor, mm	0.15
Axial AMB Lifting capacity, N	10,000
Axial gap between bearing and rotor, mm	1.0
Axial gap between catcher bearing and rotor, mm	0.5
Sensitivity of the electric eddy current sensor, mV/um	4
Switch Power Amplifier, kVA	45

The characteristic experiments, such as stiffness, damping, force-current coefficient and force-displacement coefficient have been carried out, including the 72-hours continuously running test which mainly validated the stability of the whole AMB system. Fig.13 shows the time-temperature curves of the AMB coils and Fig.14 shows the vibration of the rotor during the 72-hours running test. The bearing force experiment validated the maximum bearing capability and the disturbance excitation experiment for proved the closed-loop system's disturbance-decreasing effects.

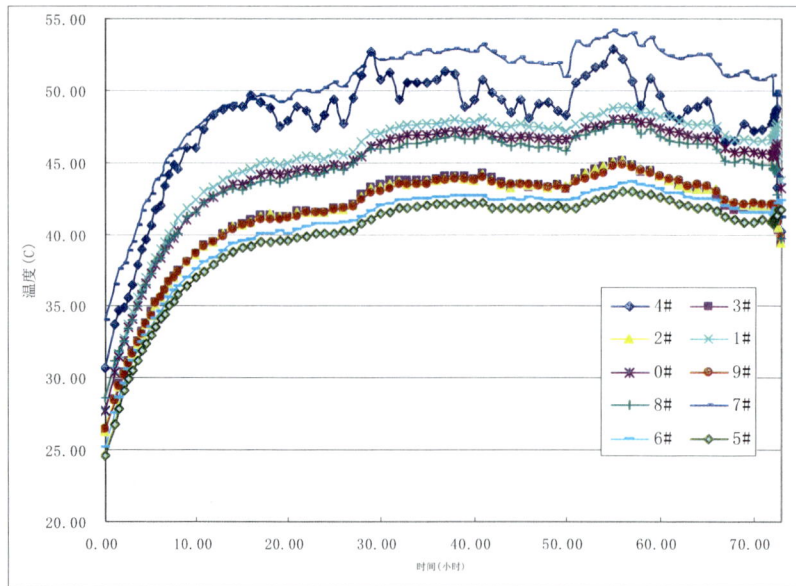

Fig.13 Time-variant Curves of the temperature nearby the magnetic bearing coils

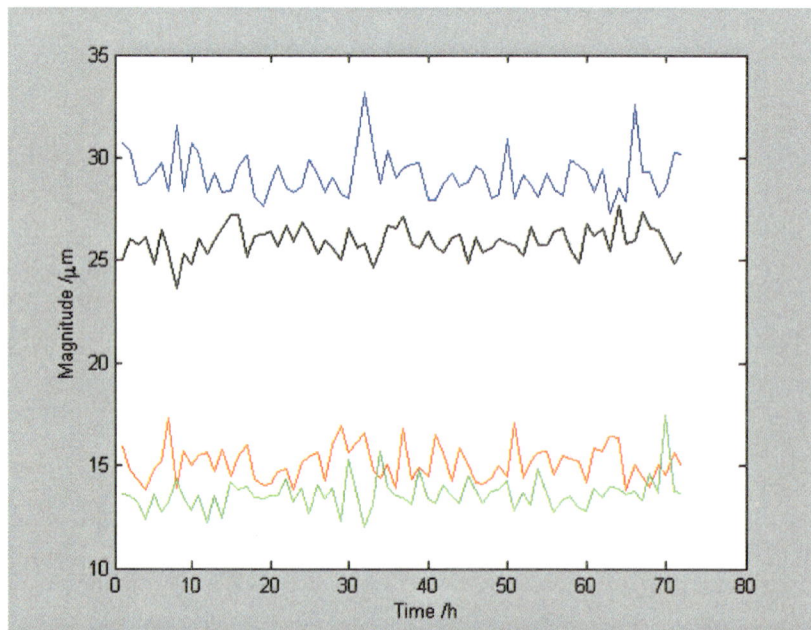

Fig.14 Maximum amplitude of radial vibration of the roto

At present, this test rig have completed more than 1000 times start-up, and fulfilled 5 times drop-down experiments fort testing the catcher bearing in 1200rpm. All the above experiments show that the technical index of the AMBs, power amplifier and catcher bearing can satisfy the design goals. This is the basis for the full scale engineering test rig.

4.4   Full scale engineering test rig

This test rig is designed for the validation of the actual engineering full size AMBs of the HTR-10GT generator rotor, shown in Fig. 15. The total length of the rotor is about 4.5m, and the weight is more than 3500kg. The main parameters are shown in Table 7. All the mechanical and electrical type, size and working principle of the AMB components are same as the future used in the actual HTR-10GT project, including the magnets, catcher bearing, DSP controller, high precision sensors, large power amplifier, cables and electrical interface.

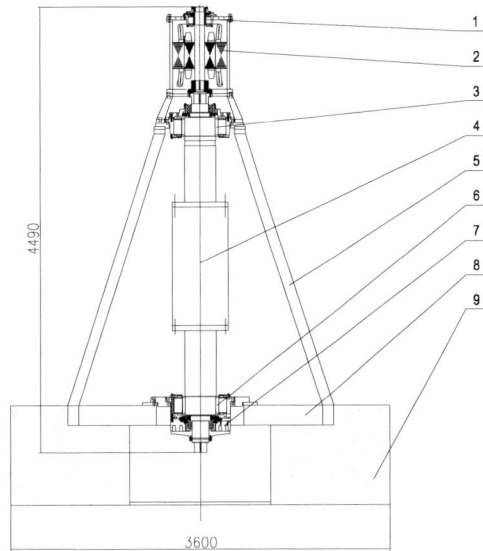

1 additive AMB; 2 driving motor; 3 upper radial AMB; 4 generator rotor;
5 supporting frameworks; 6 lower radial AMB; 7 axial AMB; 8 – metal base; 9 concrete foundation

Fig.15 System structure layout of the full size magnetic bearing test rig

TABLE 7. MAIN STRUCTURE PARAMETERS OF THE FULL SCALE ENGINEERING TEST RIG

Parameter	Value
Height / weight of the rotor, mm / kg	4,500 / 3,650
Radial AMB size, mm	Ø350×150
Radial AMB Lifting capacity, N	22,000
Radial gap between catcher bearing and rotor, mm	1.0
Axial AMB inner and outer diameter, mm	300/560
Axial AMB Lifting capacity, N	96,000
Axial gap between catcher bearing and rotor, mm	1.2
DSP Controller working frequency	75MHz
Sensitivity of the reluctance sensor, mV/um	8
IGBT Switch Power Amplifier, kVA	90
Electrical cable length, m	50

At present, the five-degree suspending and low speed running experiment is successful. Next, we will improve the control performance and increase the rotating speed up to 3600rpm in long time operation to test the properties of the whole AMB system.

5.   CONCLUSIONS

Based on the success of HTR-10, a new project of HTR-10GT which coupled gas turbine direct cycle with high temperature gas-cooled reactor has been launched. The design features and current status were presented in this paper. The arrangement of the turbomachine is selected as single shaft configuration with a gear box between turbocompressor and generator. The speed of turbocompressor is designed as 15000 r/min while the generator is 3000 r/min. The turbocompressor and generator are supported by the active magnetic bearing system. The components' R&D is in process. Among various components, the turbocompressor is considered as the most important one. Therefore, the closed air/helium wind tunnel was built and the various tests of the blade system under air and helium conditions were carried out. The active magnetic bearing (AMB) system is a key technology of the HTR-10GT project, which should support the turbocompressor and generator and control their rotors to pass through 1st and 2nd bending critical speed. Hence the INET take many efforts on the issue. Three test rigs for various magnetic bearings and rotors are built. The test results show that the AMB technology was mastered and it can be used in the HTR-10GT project.

REFERENCES

[1]   KOSTIN, V. I., et al.,. Power Conversion Unit with Direct Gas-turbine Cycle for Electric Power Generation as a Part of GT-MHR Reactor Plant. Proceedings of the 2nd International Topical Meeting on High Temperature Reactor Technology, 2004 Beijing, China.

[2]   WANG, J., HUANG, Z. Y., ZHU, S. T., Yu, S. Y.,. Design Features of Gas Turbine Power Conversion System for HTR-10GT. Proceedings of the 2nd International Topical Meeting on High Temperature Reactor Technology, 2004, Beijing, China.

[3]   Wu, Z.X., LIN, D.C., ZHONG, D. X.,. The design features of the HTR-10. Nuclear Engineering and Design **218**(1-3) (2002) 25-32.

[4]   ZHANG, Z. Y., et al.,. Design of Chinese Modular High-temperature Gas-cooled Reactor HTR-PM. Proceedings of the 2nd International Topical Meeting on High Temperature Reactor Technology, 2004, Beijing, China.

The development programme of neutron beam facilities at CARR

LIU TIANCAI, KE GUOTU

China Institute of Atomic Energy, China

Abstract: China Advanced Research Reactor（CARR）is of tank-in-pool type, cooled and moderated by light water and reflected by heavy water. The rated nuclear power is 60 MW. The maximum undisturbed thermal neutron flux is 1×10^{15} n·cm^{-2}s^{-1} in central region of reactor core and 8×10^{14} n·cm^{-2}s^{-1} in heavy water reflector. The initial criticality will be approached in 2007. There are nine horizontal channels and twenty-six vertical channels in the reactor. The application of these channels is neutron scattering (NS) experiment, radioisotopes (RIs) production, fuel and material irradiation test, neutron transmutation doping of silicon (NTD), neutron activation analysis (NAA), neutron radiography (NRG), boron neutron capture therapy (BNCT) and so on. The development programme of these facilities will be discussed in this paper.

1   INTRODUCTION

The China Advanced Research Reactor (CARR) Project, which consists of a reactor and its associated systems and experimental facilities, is a large nuclear engineering project.

CARR is located in the territory of China Institute of Atomic Energy (CIAE) ,Fangshan District of Beijing. The direct distance from CIAE to the centre of Beijing is about 37 km. The construction area of CARR is approximately 18000m^2, about 2.3 hectares of ground area occupied.

The construction period of CARR is 52 months and the initial criticality will be approached in 2007. At present, the building, reactor body, reactor engineering systems necessary for CARR operation safely, hot cell are under construction. All the utilization facilities are still in programming stage. The whole project including experimental facilities is scheduled to be completed in 2015.

The CARR Project is set up for meeting the need of science and technology development in the 21st century. Having incorporated the successful experience in construction of research reactors in China and foreign countries as reference and also with its own innovations and technical features, CARR is a safe, reliable, multipurpose research reactor of high performance. It will provide an important platform of neutron scattering (NS) experiment, radioisotopes (RIs) production, fuel and material irradiation test, neutron transmutation doping of silicon (NTD), neutron activation analysis (NAA), neutron radiography (NRG), boron neutron capture therapy (BNCT) and other application.

2   THE MAIN DESIGN PARAMETERS OF CARR

CARR is a multi-purpose, tank-in-pool, inverse neutron trap type research reactor. The reactor is cooled and moderated by light water and reflected by heavy water. The bird's eye view of CARR is shown in Fig. 1. The schematic of CARR complex structure is shown in Fig. 2. The layout of experimental channels is shown in Figure 3. The layout of reactor core is shown in Fig. 4.

The main structural and physics parameters of CARR are showed as follows:

Reactor power/MW	60
Max. undisturbed thermal neutron flux/ n·cm^{-2}s^{-1}	
in central region of reactor core	1×10^{15}
in heavy water reflector	8×10^{14}
Core height/m	0.85
Equivalent diameter of core/m	0.399
Diameter of reactor pool/m	5.5
Elevation of reactor pool water surface/m	13.2
Volume of reactor pool water/m^3	750

(Including the spent fuel pool connected)	
Inner diameter of heavy water tank/m	0.479
Outer diameter of heavy water tank/m	2.2
Height of heavy water tank/m	2.0
Material of control rod absorber	Hf
Number of safety rods	2
Number of regulating & shim rods	4
Gross worth of control rods/($\Delta k/k\%$)	36.37

Fig. 1 The bird's eye view of CARR

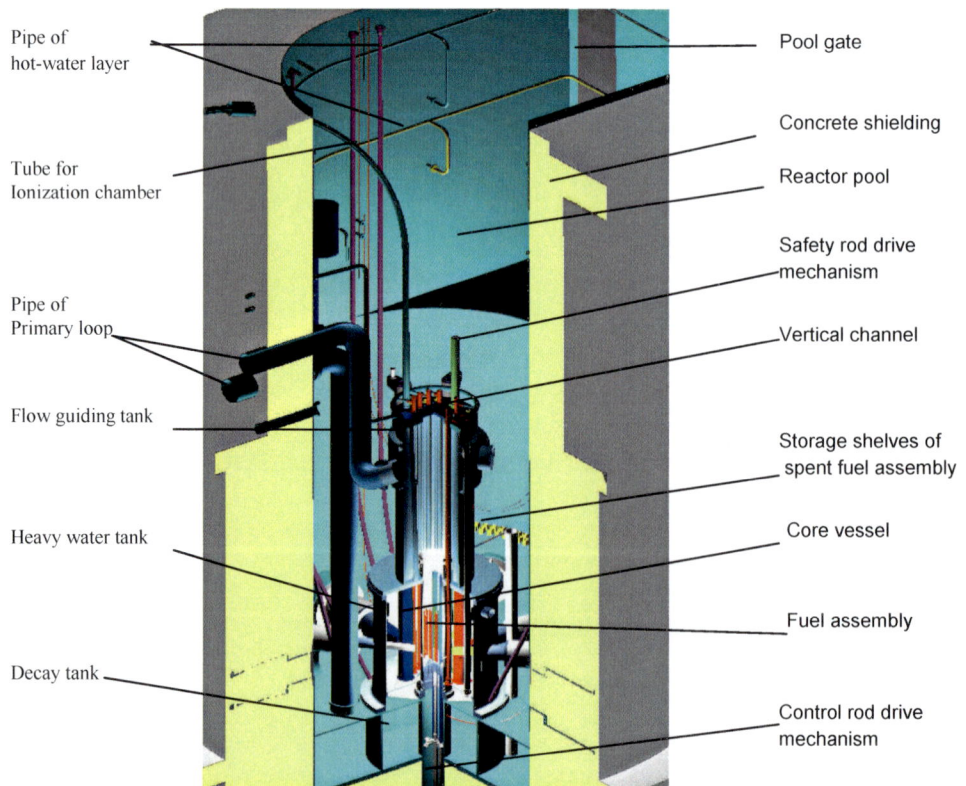

Fig. 2 The schematic of CARR complex structure

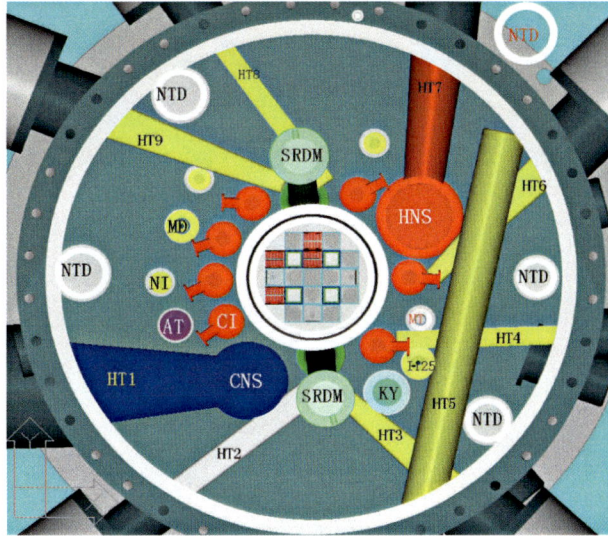

Fig. 3 The layout of experimental channels

Vertical channels	*Horizontal beam channels*
CNS: Cold neutron source guide tube	HT1: Cold neutron source beam tube
HNS: Hot neutron source guide tube	HT2: Multi-filtration neutron beam tube
MT: Material irradiation monitoring hole	HT3: Thermal neutron beam tube
I-125:Isotope I-125 hole	HT4: Thermal neutron beam tube
KY: Test loop hole	HT5: Long tangential beam tube
NTD: NTD silicon hole	HT6: Thermal neutron beam tube
MD: Mo-Tc generator hole	HT7: Hot neutron source beam tube
AT: NAA hole	HT8: Thermal neutron beam tube
NI and CI:Isotope hole	HT9: Thermal neutron beam tube
SRDM: Safety rod drive mechanism	

Horizontal beam channels	9
Vertical channels	26
Type of fuel element	Flat plate
Number of standard fuel assemblies	17
Number of follower fuel assemblies	4
Material of fuel meat	U_3Si_2 dispersed in AL
Cladding material	Aluminum alloy 6061
^{235}U enrichment (wt%)	19.75
Uranium density (g/cm^3)	4.3
Uranium for first loading/kg	55.534
Average discharge burnup/%	32.15
Average power density over reactor core (W/cm^3)	564

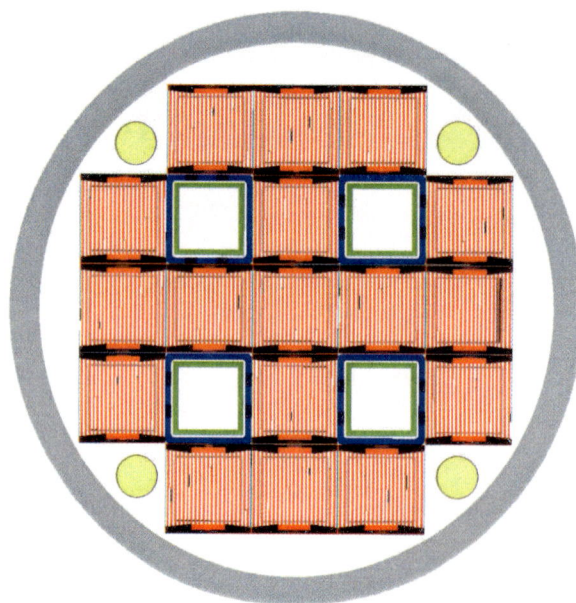

Fig. 4 The layout of reactor core

3. THE DEVELOPMENT PROGRAMME OF NEUTRON BEAM FACILITIES

3.1 Neutron scattering experiment facilities

Cold Neutron Source will play an important role in neutron scattering research. It is now of preliminary design stage at CARR. This facility is a liquid hydrogen CNS facility with helium refrigerator. CNS, HNS and some other horizontal channels will be used for neutron scattering experiments with newly installed neutron guide tubes and spectrometers. Some spectrometers such as Triple-axis spectrometer, Four Circle Diffract meter, Time of flight spectrometer, Small Angle Scattering Instrument and Powder Diffract Texture Gauge will be upgraded from HWRR (Heavy Water Research Reactor). High Resolution Powder Diffractometer, Horizontal Reflectometer, Diffraction Stress Gauge, Spin-Echo Spectrometer, Backscattering Spectrometer, Vertical Reflectometer, Cold Neutron Triple-Axis Spectrometer and High Intensity Powder Diffractometer will be developed in the future. With these facilities, CARR will provide powerful capability for conducting a great deal of fundamental and engineering applied researches covering material science, life science, envionment science, researches in physical chemistry fields and in other important relevant areas.

3.2 Radioisotope production facilities

Vertical channels with different diameters and different neutron flux levels and the automatic processing transportation systems can be used for production of radioisotopes such as ^{60}Co, ^{99}Mo, ^{131}I, ^{113}Sn, ^{125}I, ^{32}P, ^{192}Ir, ^{14}C, ^{35}S, ^{210}Po ^{160}Ho and ^{198}Au, etc. in industrial scale, which are widely used in scientific researches, radio-medicine, industry and agriculture. Various irradiation target, capsule, lead storage, handing tool, ^{125}I production loop, and ^{99}Mo/^{99m}Tc generator, etc., will be developed.

3.3 Fuel and material irradiation test facility

For meeting the need of conducting experimental researches for nuclear fuels, structural materials and component elements, specific testing facilities or systems will be built for carrying out irradiation tests.

A high temperature and high pressure testing loop with ^{3}He gas-adjusting loop will be constructed in CARR. The performance tests (including steady and transient), high burn-up test, water chemistry activity transport and corrosion test, and fuel integrity and qualification test, etc., could be conducted using this loop and relevant devices.

3.4 NTD silicon production facilities

Neutron doping for large diameter silicon ingots can be carried out by utilizing the technique of neutron transmutation. It will provide high quality semi-conductor materials for semi-conductor manufactures.

There are five vertical holes for 3", 4", 5" and 6" Si ingots in CARR. Irradiation facilities, including ingot handling equipment in and out of the pool, hydraulic drive mechanism, ingot storage rack during cooling, electric property inspecting equipment, annealing & cleaning equipment, gamma spectroscope system for fluence measurement, etc., will be developed.

3.5 NAA facilities

Fully incorporating the 40 years experience in performing neutron activation analysis at CIAE, new advanced NAA facilities will be built for conducting neutron activation analysis. This kind of analysis has reached the sensitivity up to 10^{-6}~10^{-9} gram for most chemical elements. Short-lived nuclide NAA System, Prompt Gamma Activation Analysis System (PGAA), Cold Neutron PGAA and Neutron Depth Profiling System will be developed. The facilities can be used comprehensively in industry, agriculture, medical science, environment science, geology and archaeology.

3.6 Neutron radiography facilities

Neutron radiography is a useful nondestructive testing technique that complements conventional radiography. High Resolution Static Neutron Radiography System and High-frame-rate Neutron Radiography System will be developed.

3.7 BNCT facility

Boron neutron capture therapy (BNCT) is a method of treating high-grade gliomas of the brain that involves incorporating ^{10}B into the tumor using appropriate pharmacological agents and then irradiating the tumor with thermal or epithermal neutron beams. Favorable reports on outcome have motivated considerable current research in BNCT. BNCT facility with thermal/epithermal neutron beam and associated therapy equipments will be explored at CARR.

4. CONCLUSION

The completion of CARR Project and its reliable operation will greatly enhance the capability of basic researches in the area of nuclear science and technology and reinforce the comprehensive strength of nuclear industry in China, pushing forward the development and application of nuclear technology. CIAE will sincerely welcome universities and colleges, institutes, business enterprises of domestic and foreign countries to carry out researches, development and application in different area of nuclear science and technology on CARR.

Residual stress analysis by means of neutron diffraction at research reactors — Facilities and applications at the HFR

C. OHMS, R.C. WIMPORY[+], D. NEOV, A.G. YOUTSOS[++]

Joint Research Centre, Institute for Energy, Petten, Netherlands
[+]Now: Hahn-Meitner-Institute, Berlin, Germany
[++]Now: NCSR Demokritos, Institute of Nuclear Technology — Radiation Protection, Athens, Greece

Abstract. Neutron diffraction is among the scientific techniques available at nuclear research reactors. A particular application of neutron diffraction is the analysis of residual stresses in crystalline materials. The basic principles of the method are described and the unique capabilities of the method are pointed out. The High Flux Reactor (HFR) is one of about 20 reactor based neutron sources worldwide, where instruments for this technique are available. The HFR equipment is shown and several examples of studies performed during the last decade are given. These examples include nuclear applications, investigations in automotive components and a round robin exercise executed in the context of pre-normative research on the method. The examples demonstrate that the method is also suitable for validation of computational analyses or other experimental methods. A comparison between time-of-flight instruments and monochromatic instruments and an outlook to future developments are given at the end.

1. INTRODUCTION

Research Reactors supply neutrons for a vast range of applications. In-core methods include irradiations of nuclear fuels and structural materials, production of radioisotopes for medical purposes or silicon doping. Neutron beam based methods comprise neutron scattering techniques, radiography, neutron capture therapy, etc. There are research reactors purely built for in-core applications, and there are facilities exclusively providing neutrons to neutron beams. The High Flux Reactor of the European Commission in Petten, NL, is a multi-purpose reactor predominantly used for in-core applications, but also providing neutrons to 12 horizontal beam lines.

The present paper presents one beam based application of neutrons, namely determination of residual stresses in crystalline materials by neutron diffraction.

Beam based applications have in common that the neutrons used must necessarily move – more or less – in the same direction. This is achieved by installing the testing equipment at a certain distance from the neutron source and by installing collimating equipment in the beam. As a consequence neutron beam applications operate with considerably lower neutron fluxes than in-core applications. In-core thermal neutron fluxes at bright reactor sources today range from 10^{13} to 10^{15} n/cm^{2}s [1], while instruments installed at neutron beams at such sources would only see incoming fluxes in the range of 10^8 to 10^{10} n/cm^{2}s.

Neutron scattering techniques are typical beam based techniques since here the angle between the flight paths of the incoming and scattered neutrons is measured. This would not be reasonably possible without (quasi-)parallel neutron beams.

Diffraction of neutrons is widely used to study the structure of crystalline materials [1-2] . Bragg's principle correlates the angle of diffraction with the distances between crystallographic lattice planes. This principle is also exploited for the analysis of residual stresses. Changes in lattice distance can be established by measuring the small corresponding changes of the diffraction angle. The principle is described in the next chapter to some more detail. As neutron diffraction is only one of many methods to assess residual stresses, a statement is made concerning the unique capabilities of this method.

Special equipment is needed for this technique. Although the basic principle is identical to the so-called powder diffraction, there are major differences in the types of specimens investigated, the way these are positioned in the beam and how the beam is guided to the specimen. The typical equipment used is described in section 3.

At the HFR in Petten, NL, there are two neutron diffraction facilities for residual stress determination. These facilities and their main characteristics are presented in section 4.

In chapter 5 several examples for experimental stress analyses are shown. The first example is a round robin exercise in a welded test piece, which was performed in a welded plate of ferritic steel. Another example shown here are measurements of residual strain in surface treated crankshaft sections of a car engine. These measurements are compared against numerical stress predictions for this component. The series of examples is concluded with two investigations related to nuclear applications. Residual stresses have been measured in a stainless steel plate with a single weld bead of finite length on top of it. These measurements should provide information to improve the methods for modelling of welding processes with a particular focus on repair welding. Finally, a thick bi-metallic piping weld has been investigated. This specimen was a full-scale mock-up of a reactor pressure vessel nozzle to primary piping weld.

The paper is concluded with a comparison of monochromatic neutron beam and time-of-flight (TOF) methods. A brief outlook to the future is given referring to the instruments at the most modern neutron sources in operation and under construction today.

2. PRINCIPLE OF THE METHOD

All diffraction methods for residual stress analysis are based on the Bragg principle [3]. This principle relates the lattice spacing, d, of a crystalline material to the positions of the diffraction peaks in the scattering pattern that this material generates from the incident radiation. For a given wavelength, λ of the incident radiation and a given lattice spacing, d, the Bragg equation is:

$$n . \lambda = 2\, d\, \sin\theta \qquad (1)$$

with n being an integer representing the order of the reflection observed and θ being the angle of diffraction. While the Bragg principle was established using X rays as the incident radiation, it was discovered some time later that, in accordance with the postulate made by de Broglie [4], diffraction from crystalline materials could also be obtained with small particles (electrons, protons, neutrons) as incident radiation. The wavelength in this case would be derived from the particle's momentum, p, and Planck's constant, h:

$$\lambda = \frac{h}{p} . \qquad (2)$$

It was not until well after World War II that neutrons became available for scattering and diffraction experiments in the first nuclear research reactors. From the very beginning neutron diffraction has been used for structure analysis to complement the well-known X ray diffraction techniques. However, only in the 1980's the first facilities for residual stress analysis with neutrons were developed [5].

As the presence of residual stresses changes the lattice spacing, d, of the material from its unstressed value, d_0, the measurement of this variation was used for the experimental determination of such residual stresses. The relative deformation under stress, i.e. the strain, ε, is then defined as

$$\varepsilon = \frac{d - d_0}{d_0}, \tag{3}$$

Combination of eqs. (1) and (3) renders:

$$\varepsilon = \frac{\sin \theta_0}{\sin \theta} - 1. \tag{4}$$

By measuring the scattering angle in a material element under stress, θ, and the scattering angle in an identical material without stress, θ_0, the strain can be determined in accordance with eq. (4)[1].

Residual stress and strain are magnitudes of tensor dimension and they vary with location and direction. By changing the orientation of the specimen under investigation with respect to the neutron beam one can measure strains within the specimen in different directions. By changing the position of the specimen in the neutron beam, measurements can be obtained from different locations inside a specimen. Figure 1 below illustrates this.

When strains have been measured in three mutually orthogonal directions, e.g. x, y and z, at a given location within the specimen, stresses in these directions can be determined via the generalized Hooke's law:

$$\sigma_i = \frac{E}{1+v}\varepsilon_i + \frac{vE}{(1+v)(1-2v)}(\varepsilon_x + \varepsilon_y + \varepsilon_z), \tag{5}$$

with the index i representing the measurement direction x, y or z. In the case that x, y and z coincide with the principal stress directions at the test location, the stress tensor is fully determined by measuring ε_x, ε_y and ε_z. Equation (5) is also applicable in the case x, y and z are not the principal stress directions, but in this case the off diagonal components of the stress tensor would not be 0. In most practical cases it is assumed that the specimen geometry and the nature of the production process indicate, what the principal stress directions are, and residual stress measurements are performed in accordance with eq. (5) only.

For the measurement of residual stresses there is a large number of techniques available. They can be categorised by various criteria. The diffraction techniques, for example, are non-destructive, whereas all strain gauge based techniques are considered to be destructive.

Unlike conventional X rays, neutrons can penetrate deep into most crystalline materials. This makes the neutron a non-destructive probe capable of measuring in the bulk of a material specimen under investigation. Neutron diffraction can measure stresses in 3-dimensions with a reasonable spatial resolution (1-5 mm). No other stress measurement technique offers a combination of these properties [1].

---

[1] It should be noted that the neutron wavelength, does not appear anymore in (4). This means that by this method strains can be measured without exact knowledge of the wavelength of the incident radiation. It is possible to correctly measure strain this way, without being able to exactly determine the lattice spacing.

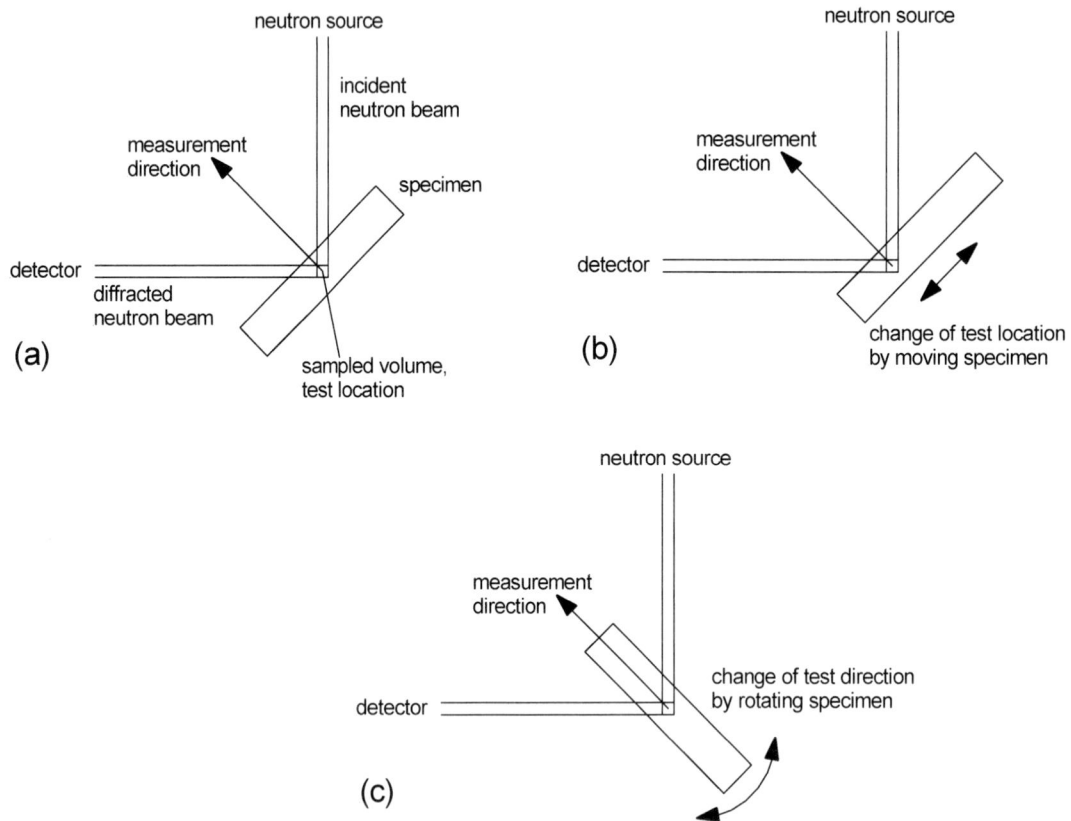

Fig. 1   Schematic of a neutron diffraction stress measurement: (a) average lattice spacing is measured over the sampled volume at the test location in a direction normal to the long side of the specimen; (b) in order to change the test location, the specimen is moved with respect to the neutron beam; (c) in order to change the measurement direction, the specimen is rotated

On the other hand it has to be noted that neutron diffraction is a very costly method, as it requires the operation of a large-scale facility – the neutron source. It cannot be disattached from the neutron source and therefore not be applied in-situ. There are limits to the thickness of components under investigation. These depend on the specimen material and on the incoming neutron flux at the facility.[2] Finally, the capability to measure stresses at surfaces is limited with neutrons, but they can measure close enough to surfaces to be an excellent complement to X ray diffraction measurements.

3. EQUIPMENT FOR MEASURING RESIDUAL STRESSES BY NEUTRON DIFFRACTION

All diffraction methods need a source of the incoming radiation. Suitable neutron sources today are research reactors and spallation neutron sources. Portable neutron sources, which do exist, cannot provide yet the required neutron flux in form of parallel beams.

Neutrons travel from the source to the instruments through the beam tube. Inside the beam tube there is normally an in-pile collimator limiting the divergence of the neutrons. In cases, where the diffractometer is installed at a considerable distance from the source (significantly more than 5 m), neutron guides would be used to transport the neutron beam to the diffractometer with limited loss of neutron flux.

---

[2] As a rule of thumb, it could be said that about 50-70 mm of steel is the best that could be achieved at the most advanced instruments today.

The neutron monochromator selects a narrow wavelength band (a single wavelength) from the incoming neutron spectrum. Monochromators are highly oriented crystals, which are used to diffract only neutrons of a single wavelength[3] in the direction of the specimen in accordance with eq. (1). Collimators can be used upstream and downstream of the monochromator in order to reduce the divergence of the neutron beam.

The specimen is mounted on the diffractometer itself. The diffractometer consists of at least five positioning stages, out of which three facilitate the linear motions of the specimen, one facilitates the rotation of the specimen around an axis normal to the plane of diffraction and the fifth stage facilitates rotation of the neutron detector around the specimen.

The neutron detector is another important piece of equipment used. Nowadays mostly position sensitive detectors are employed, so that the time-consuming scanning of single detectors around the specimen is not necessary anymore. The latest development is two-dimensional detectors, by which additional information can be obtained in terms of graininess, texture, umbrella effect etc.

The so-called neutron optical equipment is used to define the sizes of the beams impinging on the specimen and scattered in the direction of the neutron detector. At most monochromatic instruments, masks made from a neutron absorbing material (cadmium, boron, lithium or gadolinium), which have openings corresponding to the desired size of the beam, are used. Such masks would be positioned as close to the specimen as possible. Alternatively radial collimators can be used. These define the beam size at the test location more accurately from a larger distance.

The equipment described above is necessary equipment to perform residual stress analyses by neutron diffraction. Optional equipment that would be used in accordance with the necessities for a particular experiment includes:

- load frames or bending devices, when stresses/strains should be measured under external loads;
- furnaces for stress/strain measurements at elevated temperatures;
- cooling equipment for stress/strain measurement at reduced temperatures;
- Eulerian cradles for multi-directional measurements or texture analysis;
- a monitor to assess the amount of neutrons actually impinging on the specimen during an experiment;
- shielding to measure stresses in radioactive specimens, etc.

4.   RESIDUAL STRESS MEASUREMENT FACILITIES AT THE HFR

The JRC operates two residual stress diffractometers at its High Flux Reactor in Petten, NL. Both of them are monochromatic instruments employing crystal monochromators as described in chapter 3.

The Large Component Neutron Diffraction Facility (LCNDF) is installed at HFR beam tube no. 4. This facility is equipped with a large specimen positioning table, capable of carrying a mass of up to 1000 kg. In addition, the horizontal linear specimen movement ranges are – with 400 mm in the x- and y-directions – very long. The vertical positioning range is 300 mm. The neutron detector is a 32-wire multi-detector with a distance of 2 mm between adjacent wires. The total sensitive area is 63 mm wide and 127 mm high. At a distance of 1.11 m from the measurement location the detector resolution is 0.1° per wire.

---

[3] As most residual stress diffractometers at reactor sources employ monochromatic neutrons, time-of-flight methods are not considered in this chapter. A short account on these methods is given in chapter 6.

The monochromator for this facility is a pyrolytic graphite double monochromator. By using two monochromator crystals opposite to one another this monochromator facilitates selection of the neutron wavelength from a very wide range. At the LCNDF the wavelength range accessible is 0.18 to 0.6 nm

.

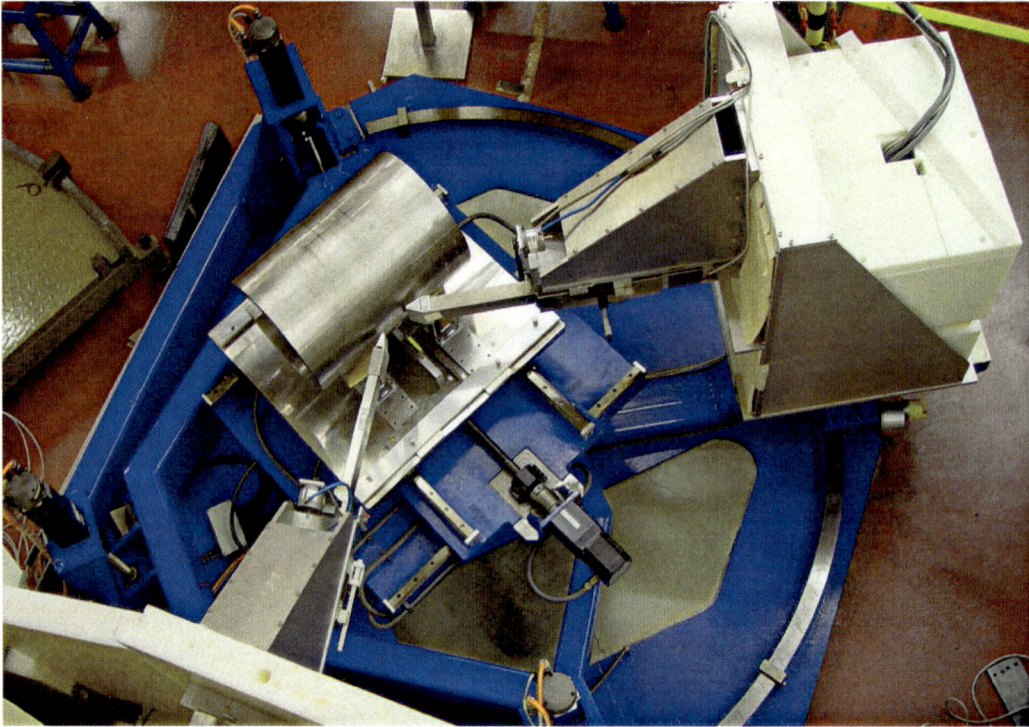

Fig. 2 The HFR Large Component Neutron Diffraction Facility

Fig. 2 shows the LCNDF during residual stress measurements in a big piping specimen with a circumferential weld. This specimen was about 0.5 m long and 0.4 m in diameter. The large positioning table can be nicely seen underneath the specimen and also the pointed beam defining masks nearly at the specimen surface are easily recognized. The neutron detector is installed in the white casing visible in the top right part of the picture.

The movement ranges and its weight capacity make this facility very well suited for investigations in larger industrial components representative of real engineering applications. Currently most investigations relate to nuclear welded components.

At beam tube no. 5 a copper monochromator is employed. Based on the use of the (111) reflection plane and a take-off angle of 76° it delivers a neutron wavelength of 0.257 nm.

The diffractometer mechanics for specimen handling and detector movement are currently being replaced. The old facility – shown in Fig. 3 – stemmed from the early days of neutron based stress measurement. It was actually a stress diffractometer derived from a pre-existing powder diffractometer. The modifications necessary to measure stresses on a powder diffractometer are simply the addition of a specimen positioning table and neutron optics to define the beam sizes. Consequently these additions had to be fitted into a limited space.

The positioning table offered a movement range of less than 100 mm in the horizontal directions and about 150 mm vertically. The maximum load on the table was not more than 20 – 30 kg.

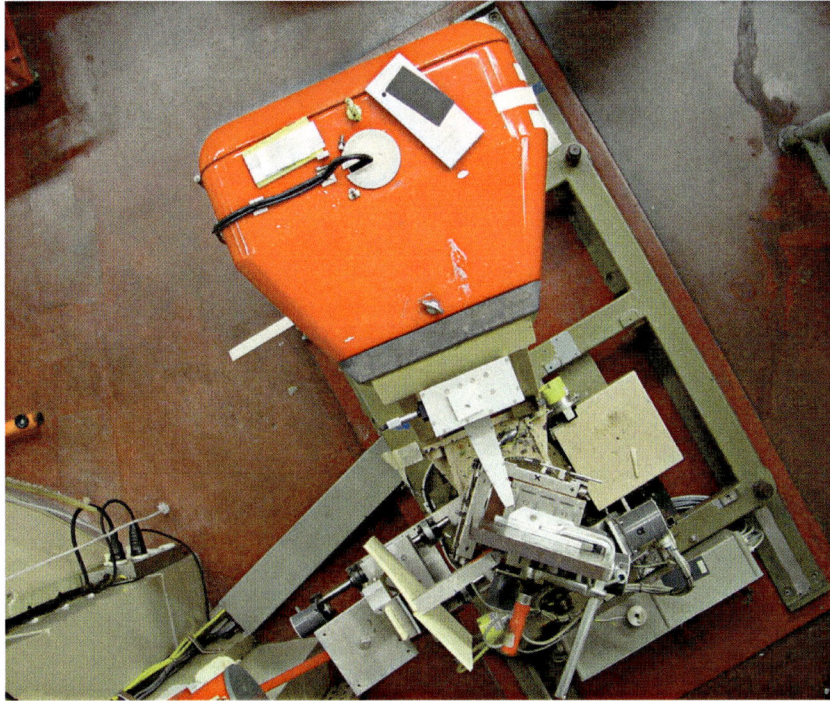

Fig. 3 The old residual stress diffractometer at beam tube no. 5

The new mechanical system is shown in Fig. 4. The movement ranges offered are 250 mm horizontally and vertically, and the load capacity is more than 200 kg. This is still somewhat smaller than the LCNDF, but the whole instrument has to fit in a more limited space. In addition to the extended capacity this instrument offers higher resolution in positioning and position encoding on all motors. The new instrument is expected to operate in 2007.

The position sensitive detector used is a 100 mm wide He^3 detector with a postion resolution of about 1 mm. At a distance of ~700 mm from the test location this corresponded to ~0.06° per pixel at the old instrument. The new diffractometer will offer the possibility to change the detector distance. Consequently the detector resolution varies. The minimum distance will be more than 700 mm.

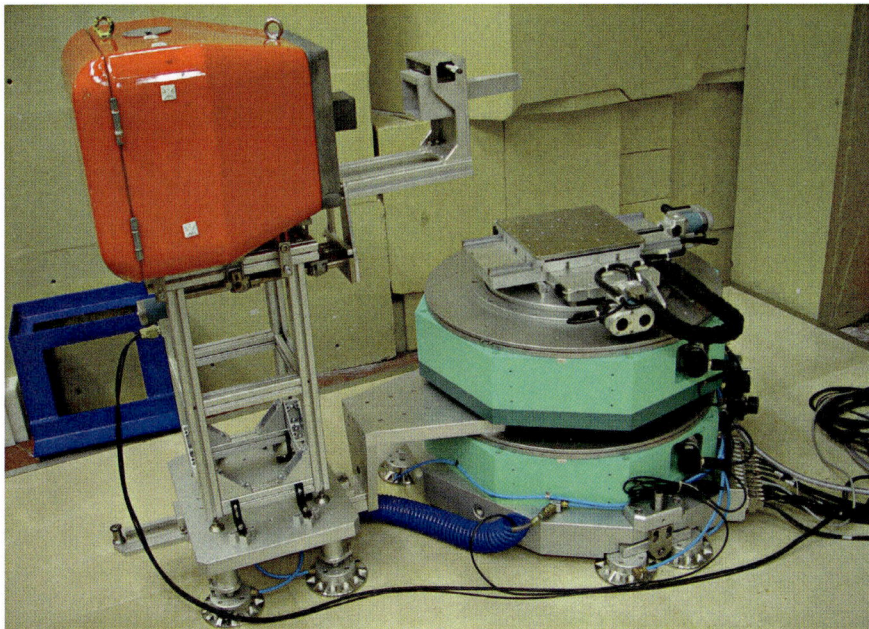

Fig. 4 The new residual stress diffractometer at beam tube no. 5

5. RESIDUAL STRESS INVESTIGATIONS BY NEUTRON DIFFRACTION AT THE HFR

5.1 Residual stress measurement round robin in a ferritic steel welded plate

In the context of the international and European pre-normative research activities VAMAS TWA 20 and RESTAND [6-8], which embarked on formulating a draft standard for the determination of residual stresses by neutron diffraction, several experimental round robin exercises were performed. For one of them, an aluminium shrink fit ring&plug, the results have been reported to some detail in [7].

Another round robin exercise executed in this context was a ferritic steel welded plate. The specimen was 12.5 mm thick, 200 mm long, 150 mm wide and had a 12 pass butt weld along its centre. The weld had been applied in a groove penetrating about three quarters through the thickness of the specimen. As can be seen from Fig. 5 below, the forces caused by the shrinking of the weld zone during cooling down, lead to substantial bending of the plate.

The purpose of the measurements was to map the residual stresses in that plate over a plane normal to the welding direction at mid-length of the specimen by measuring strains in the welding longitudinal, welding transverse and plate normal directions. A black line indicates the location of the mapping plane in the middle of the specimen in Fig. 5.

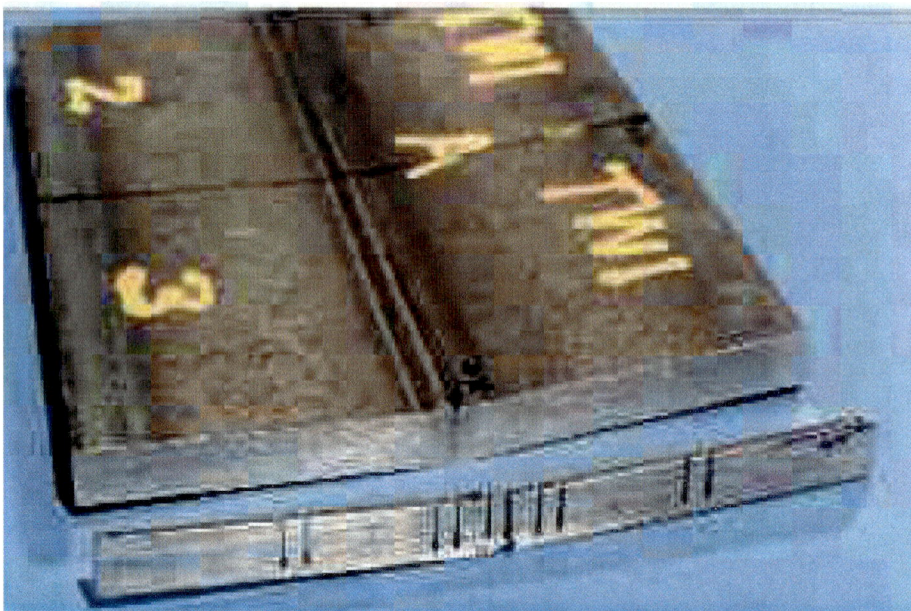

Fig. 5 Ferritic steel welded plate for round robin residual stress analysis by neutron diffraction

The measurements were done with a relatively high resolution. Sampling volumes of $2 \times 2 \times 20$ mm^3 were prescribed for measuring the welding transverse and plate normal directions, whereas for the welding direction only $2 \times 2 \times 2$ mm^3 were allowed. A very thorough mapping was performed at the HFR LCNDF with the large sampling volume, while for the welding longitudinal direction fewer positions could be measured in view of the long testing time required because of the small measurement volume.

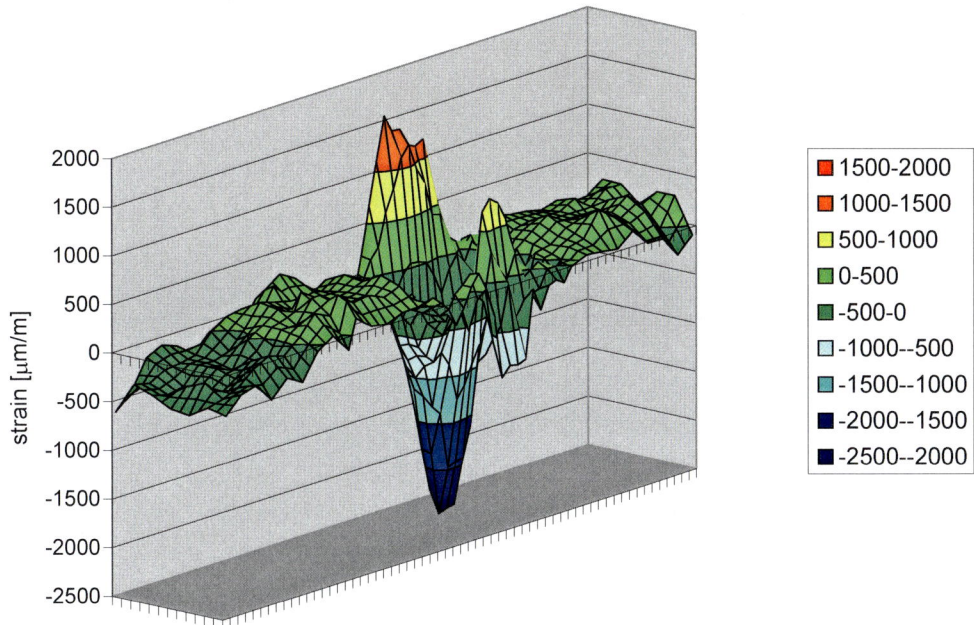

Fig. 6 Map of welding transverse strains measured at the HFR in the ferritic steel welded plate

Figure 6 shows the detailed strain map obtained this way in the welding transverse direction at the HFR. It can be seen very nicely that in the region of the weld itself the largest strains, both tensile and compressive, are found. In addition to the welding process, the subsequent bending of the plate played a major role in forming this strain/stress distribution.

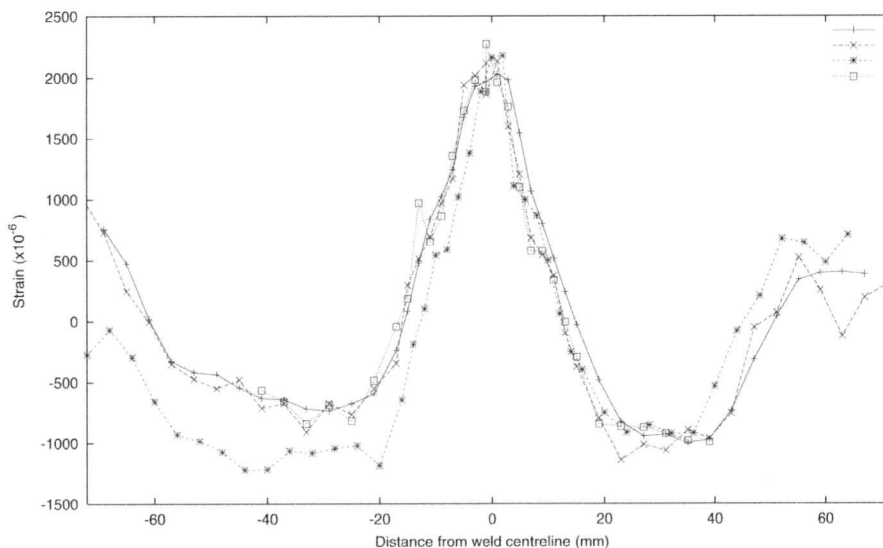

Fig. 7 Welding longitudinal strain round robin measurements obtained at 4 different neutron laboratories, incl. the JRC, from [9]

Figure 7 shows a comparison of measurements in the welding longitudinal direction obtained from 3 European laboratories, including the JRC, and one North-American laboratory. It can be seen that a remarkably good agreement between the measurements exists for 3 of the data sets, again including the JRC. During measurement of the fourth data set, which is visibly deviating from the others in terms of absolute strain, but also measurement position, several procedural mistakes had been made. The biggest one would probably be that the specimen had hit the neutron masks during the experiments, and even a small misalignment of these masks causes a shift in the measured strain, but also in the position of the measurement.

The above data demonstrate the resolution that the neutron diffraction method for stress analysis can offer and the remarkable agreement that can be achieved between facilities on nominally identical specimens, provided that experiments are executed with care and in accordance with agreed protocols.

5.2   Residual strain measurement in surface treated crankshaft sections

Crankshafts used in combustion engines are subject to severe mechanical loads during operation. In particular, wear in the stroke-bearing sections of crankshafts is of concern, because it renders the bearing sections very prone to cracking induced at the surface.

Compressive residual stresses at surfaces prevent or at least significantly delay the propagation of surface cracks, and therefore contribute to extended fatigue life of components. Several techniques can be applied to introduce surface compressive residual stresses.

The specimens selected for these investigations were made of a grey cast iron commonly used for crankshafts. The test pieces provided were bearing sections of ~5 cm diameter and ca. 8 cm length (see Fig. 8). The aim of the investigations was the determination of the combined influence of induction-hardening and deep-rolling processes on the residual stresses. The experimental results were compared against numerical predictions. The area of interest was the deep rolled groove on the edge of the bearing zone (marked "1" in Fig. 8). The expected high stress gradients necessitated the use of a small gauge volume ($2 \times 2 \times 2$ mm^3). In a component of this thickness consequently only strain measurements in two directions - axial and radial - were possible.

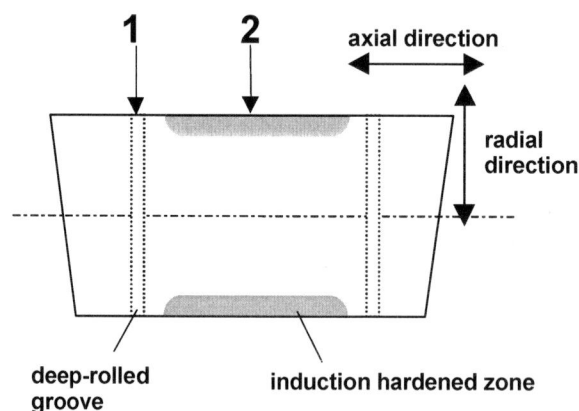

Fig. 8 Cross section of crankshaft stroke-bearing sections investigated by neutron diffraction

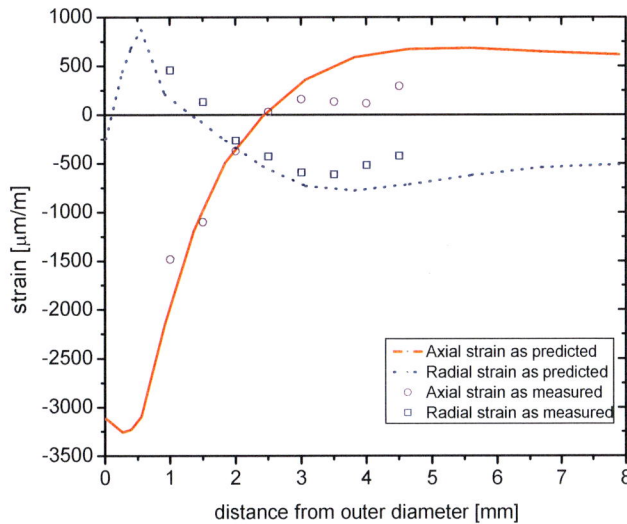

Fig. 9 Axial and radial strains in a crankshaft section subjected to deep rolling and induction hardening, comparison between HFR neutron diffraction measurements and finite element analysis performed by manufacturer, from [10]

Figure 9 shows that very good agreement between experiment and model has been achieved for a test piece subjected to a high deep rolling force plus induction hardening. It also shows that it is difficult, to obtain measurements in the near surface region by neutron diffraction, which would be highly desirable in view of the predicted near surface strains. Such results can be used tovalidate numerical predictions of manufacturing processes like deep rolling and induction hardening, which gives the engine manufacturer the possibility to optimize the manufacturing process and to minimize the probability of failure of their products.

5.3   Residual stresses in a single bead on plate stainless steel weld

The primary objective of these investigations was a thorough 3-D analysis of residual stresses around this single weld bead by both experimental and numerical means. The rationale behind this was the need for a better understanding of residual stress fields generated by weld beads of finite length. Such beads would typically be applied when it comes to repair welding. The experimental and numerical work was performed in the form of round robin campaigns in order to maximise the reliability of the residual stress data obtained.

The specimen investigated was a 17 mm thick plate of stainless steel grade 316H. The plate was 180 mm long and 120 mm wide and a single weld bead of 60 mm length had been applied on one side in the middle of the surface. Figure 10 shows a companion specimen of the same dimensions during residual stress measurement at the HFR. The specimen itself has no technical application; the experimental investigations exclusively provide information that is used to optimize the full 3-dimensional modeling procedures for finite length welding. The knowledge gained by these investigations could in the future be used to support safety cases for necessary repair welds, e.g. in the nuclear industry.

The entire experimental campaign comprised measurements of residual stresses along various lines within the specimen. One line was parallel to the weld bead, and three other lines were transverse to the bead. Those lines were located 2 mm below the welded surface of the plate. In addition three lines running through thickness from welded face to the opposite face have been selected for measurements. All strain measurements were to be performed in the welding longitudinal, welding transverse and plate normal directions, so that residual stresses could be derived from these measurements in accordance with Eq. (5). At the HFR measurements have been performed at beam tube no. 5 using a scattering volume of $3 \times 3 \times 2$ mm^3. For the calculation of the residual stresses, diffraction elastic constants found in the literature [11] for this austenitic steel and the (111) diffraction plane were used. Figure 11 depicts the longitudinal residual stresses measured along a line transverse to the weld 2 mm below the welded surface. As can be seen the measurements performed cover only about half of the actual width of the specimen. This is mainly related to the limited movement range of the positioning stages at the old diffractometer (see section 4).

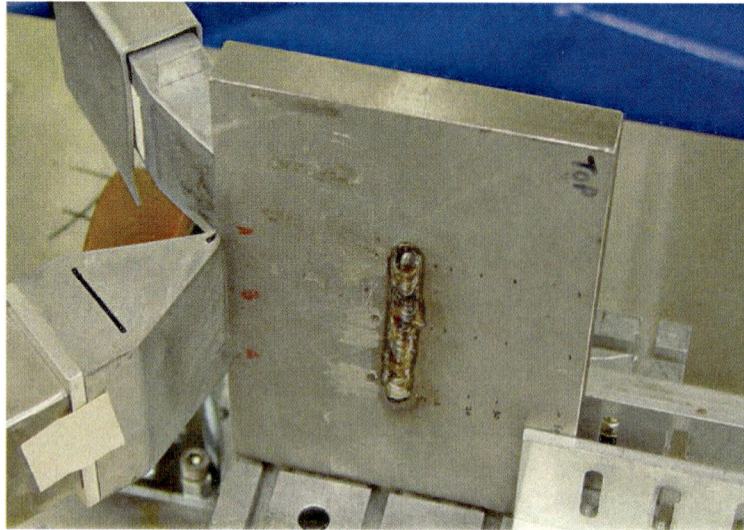

Fig. 10 Single bead weld on plate specimen during residual stress measurements at the HFR

The results show very high tensile residual stresses of up to 350 MPa in the weld region. Part of the compressive residual stresses that must be present in this cross section of the specimen in order to balance the tension in the weld zone is shown on the left side in Fig. 11. The transition from tension to compression was found to be 20-25 mm from the centre of the weld bead. One must keep in mind that the test locations shown do not cover an entire cross section of the specimen, but only a line. Therefore the stresses found along this line do not necessarily have to balance, which indeed they appear not to do.

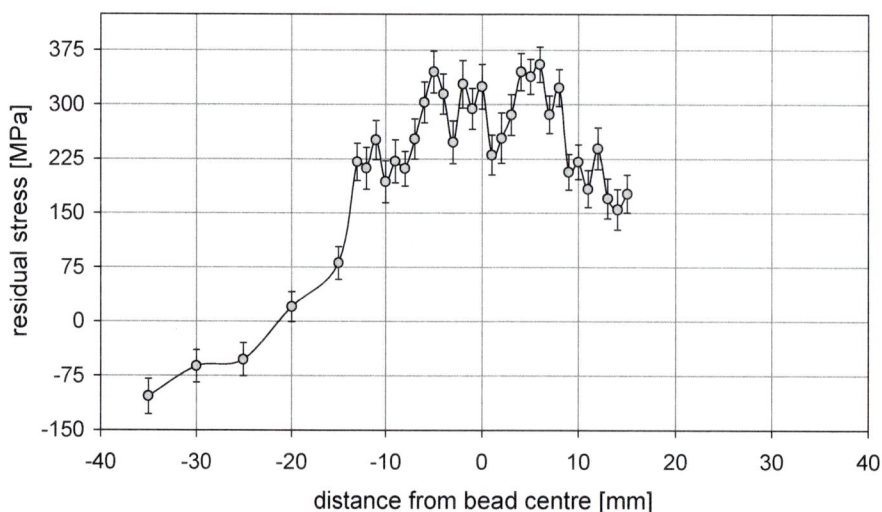

Fig. 11 Welding longitudinal residual stresses in a single bead on plate weld

5.4 Residual stress measurement in a bi-metallic steel piping girth weld

Connection of nuclear pressure vessels made of low alloy steels to piping systems made of stainless steel involves the application of dissimilar metal girth welding. This process generates a stress pattern deviating from that of a monolithic piping weld. In addition to the thermo-mechanical processes caused by the addition of extremely hot material, there is the mismatch of the thermo-mechanical properties that contributes to the formation of residual stresses in such a case. This mismatch would cause significant residual stresses to be present even after a thermal heat relief treatment.

The specimen under investigation in this case was a full-scale mock-up of such a pressure vessel nozzle (ferritic steel grade A508) to primary piping (stainless steel grade 316L) weld. The main objective of these measurements was to provide experimental data for validation of complex numerical models used to assess the residual stresses. Figure 12 shows the mock-up during measurement of circumferential strains at the LCNDF. The specimen was 500 mm long, about 450 mm in diameter and had a wall thickness of 50 mm. With such a wall thickness measurements were only possible with a very large sampling volume of $10 \times 6 \times 10$ mm^3 for the hoop direction, and this thickness proved to be at the limit of the possibilities at the HFR.

Measurements were performed in the weld region of the component. Because of less favorable texture in the piping axial and radial directions, at many locations in the weld pool only one or two measurement directions could be investigated, and here determination of stresses in accordance with (5) was impossible. Figure 13 shows the comparison of hoop strain measurements only, which were taken at various levels of depth below the outer surface of the tube, with the result of a numerical assessment of this welding process (A detailed account on the analysis procedures is given in [12].). The comparison between measurement and model is remarkably good near the outer surface of the pipe, but deteriorates toward the inner surface. This illustrates, how difficult such numerical analyses of multi-pass welding processes still are today, and how important the availability of reliable experimental techniques is for the validation of such complicated numerical analyses.

Fig. 12 Bi-metallic piping weld specimen during residual stress measurement at the LCNDF; the material interface can be seen at the specimen surface

6. THE TIME-OF-FLIGHT TECHNIQUE

In sections 2 to 5 the method of residual stress measurements by neutron diffraction has been presented based on the principle of monochromatic instruments. Equations (3) and (4) are in this case the basis for the measurement of strain. The alternative time-of-flight technique is not given a lot of room here because it is almost never applied at reactor based instruments. Nevertheless for the sake of completeness the principle is outlined.

In accordance with the de Broglie principle (2) the wavelength of the neutron is inversely proportionate to its momentum, and consequently to its travel velocity, v:

$$\lambda \sim \frac{1}{v}. \tag{6}$$

When considering the length of the flight path, L, and the time, τ, the neutron needs to travel the distance L, the resulting relation is:

$$\lambda \sim d \sim \tau. \tag{7}$$

Combination of (7) and (3) renders:

$$\varepsilon = \frac{\tau - \tau_0}{\tau_0}, \tag{8}$$

which means that by measuring the time-of-flight that neutrons from the source need to reach the neutron detector after diffraction from a material specimen, it is possible to analyse strains and stresses in a very similar way as it is done at monochromatic instruments. In order to facilitate measuring the flight time it is necessary to have a pulsed beam available mostly at spallation neutron sources.

The advantage of the method is that the entire thermal neutron spectrum can be used. Hence, it allows simultaneous measurement of multiple crystallographic planes using the same detector, which is not possible at monochromatic instruments.

As stated before, this principle is almost never applied at research reactors. The newest reactor based residual stress diffractometers at Insitut Laue Langevin (ILL), France [13], FRM-II in Germany [14] and OPAL in Australia [15] operate with neutron monochromators. The main reason for this is the loss of neutron flux when a steady state beam is converted into a pulsed beam, which requires blocking the beam for a very large percentage of the available measurement time.

7. RESIDUAL STRESS MEASUREMENTS BY NEUTRON DIFFRACTION TODAY AND TOMORROW

The method of residual stress measurement by neutron diffraction has been explained. A number of examples for investigations performed at the High Flux Reactor of the European Commission's Joint Research Centre have been given. This demonstrated the potential of the technique even when taking into account that the HFR facilities are not at the top of the class today in terms of neutron flux on the specimen.

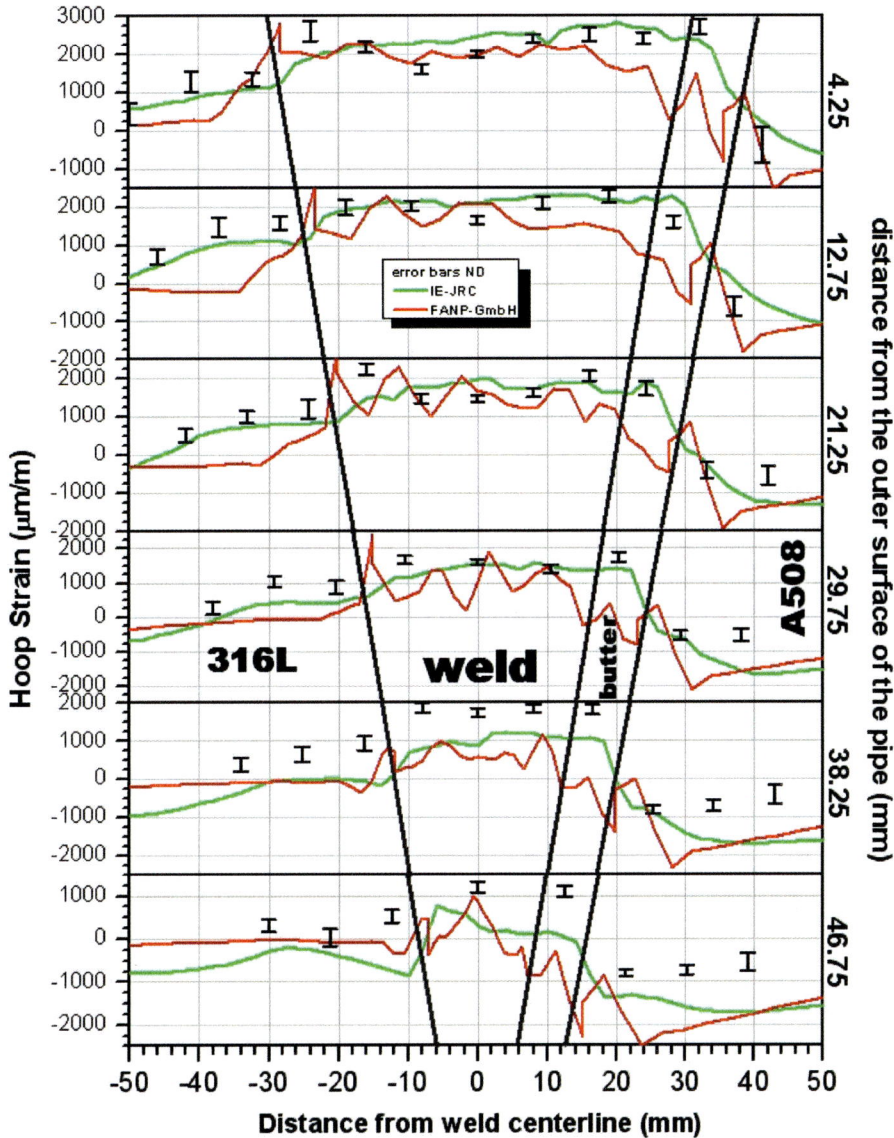

Fig. 13 Circumferential strains in bi-metallic piping weld; comparison of neutron diffraction data with numerical analysis

The best reactor based facilities today are located at the ILL, the FRM-II and at OPAL. These facilities have been optimized for the most important requirements in neutron based residual stress analysis: neutron flux, measurement resolution and flexibility in choice of wavelength and specimen size. These instruments demonstrate the tremendous progress that has been achieved since the early days of the method in the 1980's.

However, the reactor as the source of neutrons for beam based research has probably reached its limit already. Significant improvements in neutron flux offered by the source have not been made for almost 40 years. Nowadays significant source flux improvements can only be achieved with the development of high performance spallation sources. The United States and Japan are already in the process of construction of new sources, while Europe having developed the technical and scientific cases for such a project has not decided on the realization yet.

Although research reactors will not be on the leading edge of neutron beam science in the years to come, there is a substantial need for the existence of reactors as regional research centers. Over the next 20 years a significant decrease in numbers is going to become reality, if no replacement projects are materialized[4]. In the long run this is also going to have an impact on the output of the top facilities, as lack of regional centers will lead to a lack of scientists and engineers performing work at the best spallation sources.

The newest facilities already in operation [13-15], and the facilities that are going to come online in the United States [16] and in Japan promise to open new possibilities in neutron beam stress analysis. One could imagine that steel specimens of 70 mm thickness or more could be investigated in the future. On the other end of the size scale it is going to be possible to do investigations with sampling volumes significantly smaller than 1 mm^3. Strain measurement in single grains of a polycrystalline aggregate might become feasible.

Last but not least it is already foreseen to study dynamic processes on residual stress diffractometers. Phenomena, such as crack propagation are already investigated in a quasi-dynamic mode.

8. CONCLUSIONS

The latest developments in neutron stress measurement technology open fascinating new possibilities for scientific and engineering studies in the area. On the other hand a future shortage of regional neutron centers must be expected, which is going to have a negative impact on the amount of scientists and engineers being able to get access to neutron diffraction in the future.

ACKNOWLEDGEMENTS

All JRC work presented in this paper has been executed within the European Commission's Framework Programmes for Research, Technological Development and Demonstration. The authors wish to express their gratitude to their colleagues who made significant contributions to the work presented, in particular D.E. Katsareas, R.C. Wimpory, P. Hornak, T. Timke and P. van den Idsert.

In addition we would like to acknowledge the contribution of the partners within the various projects, from which work was presented in this paper, i.e. NET, RESTAND, VAMAS TWA20, ADIMEW and ENPOWER.

REFERENCES

[1]   HUTCHINGS, M.T., WITHERS, P.J., HOLDEN, T.M., LORENTZEN, T., Introduction to the Characterization of Residual Stresses by Neutron Diffraction, Taylor & Francis, Boca Raton, London, New York, Singapore, ISBN 0-415-31000-8 (2005)

[2]   KEARLEY, G.J., BOUWMAN, W.G., van WELL, A.A., VISSER, D., RAMZI, A., (Eds.), The Orange Book – Dutch Neutron and Muon Science 2000-2003, Netherlands Organization for Scientific Research, Dutch Neutron Scattering Society and Interfaculty Reactor Institute of the Delft University of Technology, ISBN 90-70608-92-8, (2003),

---

[4] It appears at the moment that more than 50% of the European and North-American reactor sources are going to be gone without replacement 20 years from now.

[3]   BRAGG, W.L., "The Diffraction of Short Electromagnetic Waves by a Crystal", Proceedings of the Cambridge Philosophical Society, **17** (1912) 43–57

[4]   de BROGLIE, L., PhD thesis, 1924

[5]   ALLEN, A.J., HUTCHINGS, M.T., WINDSOR, C.G., ANDREANI C., Advances in Physics 34, No. 4 (1985) 445-473.

[6]   WEBSTER, G.A., YOUTSOS, A.G., OHMS, C., WIMPORY, R.C., in: Recent Advances in Experimental Mechanics - in Honor of Prof. Isaac M. Daniel, Kluwer Academic Publishers, Dordrecht, ISBN 1-4020-0683-7, (2002) 467-476

[7]   WEBSTER, G.A. (Ed.), Neutron Diffraction Measurements of Residual Stresses in a Shrink-Fit Ring and Plug, VAMAS Report No. 38, ISSN 1016-2186, (2000)

[8]   WEBSTER, G.A., WIMPORY, R.C., (Eds.), ISO-Technology Trends Assessment: Polycrystalline Materials - Determination of Residual Stresses by Neutron Diffraction, Reference No. ISO/TTA 3 2001(E), ISO, Geneva, (2001)

[9]   HUGHES, D.J., WEBSTER, P.J., MILLS, G., Materials Science Forum, **404-407**, (2002) 561-566

[10]   YOUTSOS, A.G. (Ed.), Residual Stress Standard using Neutron Diffraction (RESTAND), Final Report, Technical Annex, Joint Research Centre, April 2002

[11]   EIGENMANN, B., MACHERAUCH, E., Mat.-wiss. u. Werkstofftechnik, **27** (1996) 426-437

[12]   KATSAREAS, D.E., OHMS, C., YOUTSOS, A.G., Proceedings of ASME/JSME Pressure Vessels and Piping Conference 2004, PVP-Vol. 477, M.A. PORTER, T. SATO (Eds.), ASME, New York, ISBN 0-7918-4672-5, (2004) 29-37

[13]   PIRLING, T., BRUNO, G., WITHERS, P.J., Materials Science Forum, Vol. **524-525** (2006) 217-222

[14]   HOFMANN , M., et al, Materials Science Forum, **524-525** (2006) 211-216

[15]   KIRSTEIN, O., Journal of Neutron Research, **11**, No. 4 (2003) 283-287

[16]   http://www.sns.gov

The structural materials of tritium-breeding blankets in fusion reactors

J.-L. BOUTARD

EFDA-CSU Garching, Germany

Abstract: The basic principles of a fusion reactor based on magnetic confinement in a Tokamak configuration are schematically given. Elements of design of T-Breeding Blankets are presented. For the dual-coolant Tritium (T)-breeding concept elements of design are detailed with the operating conditions and the selection of structural materials. The contribution of Small Angle Neutron Scattering (SANS) to characterise point defect & He accumulation resulting from He implantation used to simulate fusion reactor conditions is highlighted. Finally the role of a realistic multi-scale modelling to develop radiation resistant materials under fusion reactor relevant conditions is sketched. The need for increasing the sensitivity of SANS for sampling smaller specimens irradiated either with ion beam or the 14 MeV source in IFMIF is stressed.

1. BASIC PRINCIPLES OF FUSION REACTORS

In a fusion power plant the energy will be produced by the fusion reaction between Deuterium (D) and Tritium (T) nuclei:

$$D + T \rightarrow {}^4He(3.56MeV) + n(14.03MeV)$$

As a consequence 80 % of the fusion energy is carried by the 14 MeV neutrons and 20% by the 3.6 MeV alpha particles. In order to have a significant effective cross section for this reaction the deuterium and tritium nuclei must have a sufficiently high kinetic energy or temperature of the order of 100×10^6 eV. In fusion reactors such kinetic energies will be obtained in magnetically confined hot plasmas. In addition for a fusion power plant the thermo-nuclear plasma shall meet the condition of self-sustained ignition i.e. have an energy relaxation time τ_E , an electronic density n and a temperature T high enough so that n.T.τ_E >5 x $10^{21} m^{-3}$ keV.s .

The magnetic configuration to meet these conditions in fusion power plants will be most probably of Tokamak type, as in most of the present physics machines and for ITER. An example of such a torus configuration is shown in figure 1. Two in-vessel components are particularly important: the divertor and the tritium-breeding blankets. The divertor is essential for a self-sustained ignition: (i) it purifies the plasma by locally creating a magnetic configuration to evacuate the alpha particles and impurities, and, (ii) it extracts 20% of the produced fusion energy. The tritium breeding blanket requirements are three-folds. They first extract the energy from the 14 MeV neutrons. Secondly as the half life time of the tritium being only ~12 years tritium does almost not exist on the earth and has to be produced within the blanket using the D-T fusion neutrons and the following nuclear reaction:

$$n + {}^6Li \rightarrow T + {}^4He \quad + 4.78MeV$$

For tritium self-sufficiency every D-T fusion neutron has to be used to produce at least one tritium atom. Due to unavoidable losses tritium-breeding blankets also incorporate neutron multiplier functional materials such as Pb or Be. Last but not least tritium-breeding blankets are to provide the shielding of the vacuum vessel and of the superconducting magnets of the Tokomak.

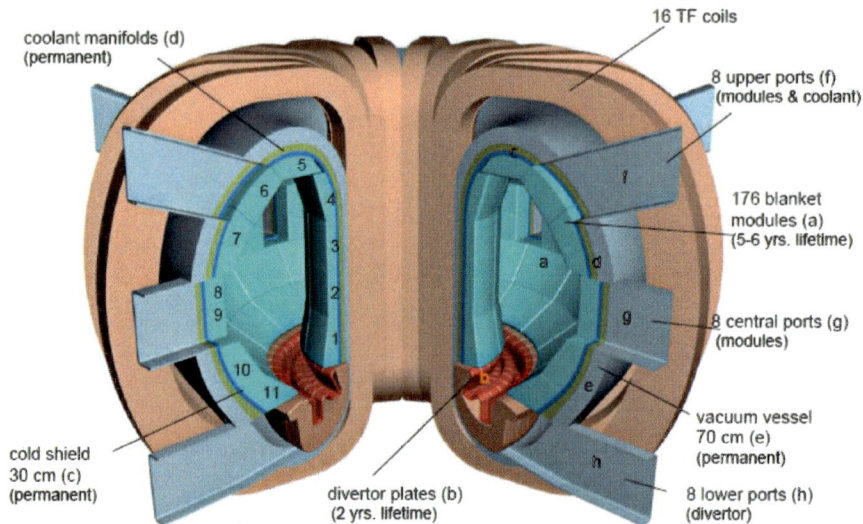

Fig. 1: Schematic view of a Tokamak fusion reactor [1]

2. TRITIUM-BREEDING BLANKET CONCEPTS, OPERATING CONDITIONS AND STRUCTURAL MATERIALS

Several concepts are being contemplated for the tritium breeding blankets based on (i) solid lithium ceramics cooled by either water or helium, and, (ii) liquid lead lithium eutectic within a dual-coolant or He-cooled concept. The dual-coolant concept of tritium breeding , and the operational parameters are given in Fig.2 and Table 1 respectively.

Within the Pb-Li eutectic the Pb nuclei act as neutron multiplier atoms, the Li nuclei provide the tritium-breeding. In the dual coolant concept the eutectic itself contributes also to the cooling of the blanket. The foreseen structural materials are Reduced Activation Ferritic-Martensitic (RAFM) steels with an upper operation temperature limit of 550^0C. Oxide Dispersion Strengthened (ODS) ferritic-martensitic steel is used as a functional protection layer of the First Wall, faces directly the plasma and can be used up to temperatures of 700^0C. Inserts made of SiC-SiC composites are used in the Pb-Li channels to (i) thermally insulate and protect the martensitic steels from too high operating temperatures, and, (ii) act as low electrical conductivity material to limit electrical current loops through the metallic structure and thus mitigate the Magneto-Hydrodynamic effects within the liquid metal Pb-Li.

The structural materials will be submitted to high radiation effects due to the 14 MeV neutrons resulting into 30 to 80 dpa or even 100 to 150 dpa for the first Demonstration or Prototype reactor, respectively [2]. Simultaneously the 14 MeV neutrons will produce significant concentrations of He and H, respectively ~10 and ~ 45 appm/dpa for the parts close to the First Wall in front of the plasma.

These operating conditions were determinant in the selection of the 9% Cr martensitic steels as structural materials since traditional 9%Cr-1Mo or modified 9%Cr-1Mo have not exhibited swelling up to doses higher than 100 dpa and shown very low embrittlement under irradiation within the sub-assemblies of the Fast Breeder Reactors in the temperature range 400 to 550^0 C [3]. For fusion reactors tailored chemical compositions were developed to have Reduced Activation (RA) materials that can be recycled hands-on after ~100 years following the end of life of the components. For this purpose the carbide former elements such as Mo or Nb are replaced by W and Ta. The developed RA 9%Cr-V-W-Ta martensitic steels [4] have mechanical and physical properties very similar to the classical modified 9%Cr-1Mo martensitic steels commonly used in boilers and steam generators in conventional power plants. Similarly the modern stoichiometric SiC fibers with cubic crystalline structure have shown an excellent stability under irradiation in Materials Testing Reactors (MTR) [5].

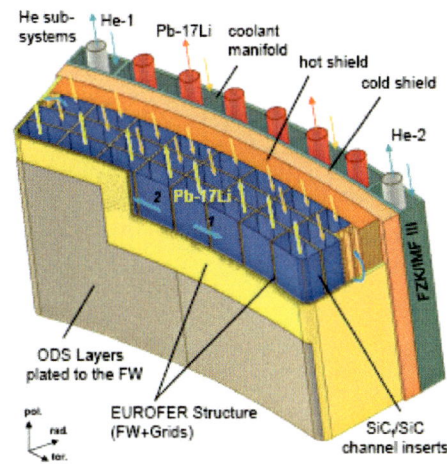

Fig. 2: Schematic view of the dual-coolant concept of tritium breeding blanket with the selection of structural and functional materials [1]

TABLE 1. OPERATING TEMPERATURE RANGES, PRESSURE AND INLET & OUTLET TEMPERATURES OF THE HE COOLANT AND OF THE PB-LI EUTECTIC [1]

Fusion Power Reactor Dual-Coolant T-Blanket	
He, 80 bars	**Pb-17Li**, ~bar
300, 480 ^{0}C	480-700 ^{0}C
Dual-Coolant T-Blanket	
Martensitic Steels (550 ^{0}C)	
ODS Ferritic steels (700 ^{0}C)	
SiCf-SiC th. & elect. insulator	
F W: T max= 625 ^{0}C	
Channel: Tmax= 500 ^{0}C	
Insert: Tmax~1000 ^{0} C	
Dose max: ~100 dpa	

3. LOW TEMPERATURE EMBRITTLEMENT OF MARTENSITIC STEELS UNDER FUSION REACTOR RELEVANT CONDITIONS

Irradiation under fusion reactor relevant conditions affects the three main parameters, which are defining the in-service properties: (i) the crystalline structure is disturbed by the production of atomic displacements, (ii) the production of He and H modifies the chemical composition and (iii) the microstructure is significantly changed by the nucleation and growth of a dense population of He-vacancy clusters that will deeply affect the mechanical properties. Characterising this microstructure and the associated impairing of the mechanical properties is essential in the development of Fusion resistant materials.

Due to the lack of a 14 MeV neutron source several methods or tricks have been used to simulate the accumulation of point defects and He under fusion reactor relevant conditions (see [6] for an overview):

- Boron or nickel doped steels were fabricated and irradiated in MTR. Nevertheless these methods give questionable data. Due to its low solubility boron is well known to segregate very easily to grain-boundaries so that the He produced by the reaction $^{10}B(n,\alpha)^{11}B$ is not uniformly distributed. With Ni-doping, the two-stage reaction ^{58}Ni $(n_{th},\gamma)\rightarrow$ $^{59}Ni(n_{th},\alpha)$ is producing He. Nevertheless nickel strongly modifies the martensite transformation points and thus the microstructure & mechanical properties of the doped materials in their initial metallurgical condition.

- ^{54} Fe doped martensitic steels are being fabricated. Helium should be produced uniformly within the materials irradiated in MTR but with a He/dpa ratio ~2 appm/dpa. First results have been reported in [7].

- Another method which has been extensively used is the uniform implantation of energetic α particles of energy in the range 20 to 100 MeV produced by cyclotron facilities. Using an energy degrader the He atoms can be uniformly implanted on thicknesses ~100 μm allowing tensile tests and microstructure examination via Transmission Electron Microscopy (TEM) and Small Angle Neutron Scattering (SANS). The He/dpa ratio is typically ~10 000appm/dpa.

As shown in fig. 3 below the implantation of Heat 250^0C clearly results in an important hardening and significant loss of ductility. TEM examination of samples implanted at 250 and 550^0C consists in He clusters or bubbles as shown in figure 3 (c) and (d) [9].

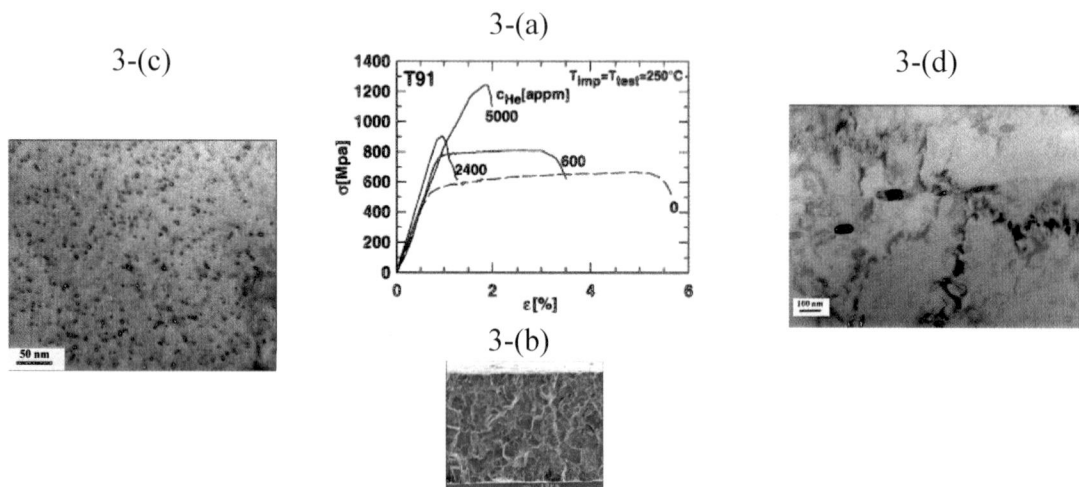

Fig. 3: (a) Tensile behaviour of modified 9%Cr-1Mo (T91) after uniform implantation of α-particles of 23 MeV up to 5000 appm at 250^0C carried out at the compact cyclotron of Forschungszentrum Jülich (FZJ) [8]. (b) Intergranular fracture surface after implantation at 250 ^{0}C along the prior austenite grains observed by Scanning Electron Microscopy (SEM). (c) TEM observation of He-V nano-clusters in samples implanted at 250^0 C up to 5000 appm He [9].(d) TEM observation of He-V bubbles along dislocations and interfaces in samples implanted at 550 ^{0} C up to 5000 appm He [9]. For 5000 appm He the dose is ~0.8dpa.

4. SANS: AN EFFECTIVE TECHNIQUE TO CHARACTERISE MICROSTRUCTURE AND SAMPLE MACROSCOPIC VOLUMES.

Cold neutrons, generally used at the temperature of the liquid hydrogen, have wave-lengths greater than the Bragg diffraction conditions and can be scattered by nano-metric precipitates, provided that these precipitates introduce nuclear or magnetic heterogeneities. The differential cross section of the scattered neutron intensity is the following:

$$\frac{d\Sigma}{d\Omega} = \sum_{p=1}^{N_p} (\Delta\rho_{nucl}^2 + \Delta\rho_{mag}^2 \sin^2 \alpha) V_p^2 F_p^2(q)$$

where $\Delta\rho_{nucl}^2$ and $\Delta\rho_{mag}^2$ are the nuclear and magnetic contrasts between precipitates and matrix, α is the angle between the magnetic field and the scattering vector $\vec{q}$ defined as the difference between the scattered wave vector $\vec{k}'$ and the incident one $\vec{k}$. V_p and $F_p^2(q)$ are the volume and the form factor of the precipitates respectively. The summation is on the precipitates population sampled by the neutron beam.

The nuclear and magnetic contrasts are given by the following formulas:

$$\Delta\rho_{nuc} = b_{nuc}^m / \Omega_m - b_{nuc}^p / \Omega_X \qquad\qquad \Delta\rho_{mag} = b_{mag}^m / \Omega_m - b_{mag}^p / \Omega_X$$

where b_{nuc}^m, b_{mag}^m, b_{nuc}^p, b_{mag}^p are the coherency length of the nuclear (nuc) and magnetic (mag) scattering in the matrix (m) and in the precipitates (p). Ω_m, and Ω_X are the atomic volumes in the matrix and in the precipitates, respectively.

Therefore the scattered intensity depends on the chemical composition of the precipitates and on the corresponding atomic volume: assumptions are to be made or information from other techniques is needed to analyse the scattered intensity.

Assuming that the He pressure within He-clusters is the equilibrium one and the clusters are spherical, the analysis of the total intensity scattered by RA martensitic steels uniformily implanted up to 400 appm He at 250 ^{0}C using the 104 MeV α particles produced by the cyclotron of Forchungszentrum Karlsruhe (FSJ) results in a size distribution in good agreement with the one determined by TEM [10]. Moreover the SANS technique allows to follow the thermal annealing of the He-cluster population at 525, 825 and 975 ^{0} C [10], these temperatures being typical of the one experienced by Heat Affected Zone (HAZ) during welding or re-welding operation.

In the case of He-clusters, the magnetic coherency length is equal to zero due to the fully occupied 2s shell of He atoms, which is non – magnetic. Therefore the maximum magnetic SANS intensity obtained by the difference of the intensity scattered parallel and normal to the magnetic field:

$$\left(\frac{d\Sigma}{d\Omega}\right)_{mag} = \left(\frac{d\Sigma}{d\Omega}\right)_{parallel} - \left(\frac{d\Sigma}{d\Omega}\right)_{normal} = \sum_{b=1}^{N_b} \Delta\rho_{mag}^2 V_b^2 F_b^2(q)$$

is independent from the atomic volume and thus from the He pressure.

Assuming (i) spherical He-clusters in agreement with TEM observations to calculate V_b and $F_b(q)$, and, (ii) a mono-modal Gaussian size distribution, the magnetic signal scattered by the microstructure shown in figure 3-(c) and 3-(d), was analysed. The obtained size distributions are given in figure 4. The main parameters, mean radius, standard deviation and bubble density obtained by TEM and SANS are compared in the table 2 for 550 ^{0}C. The size distributions determined by SANS and TEM are in very good agreement.

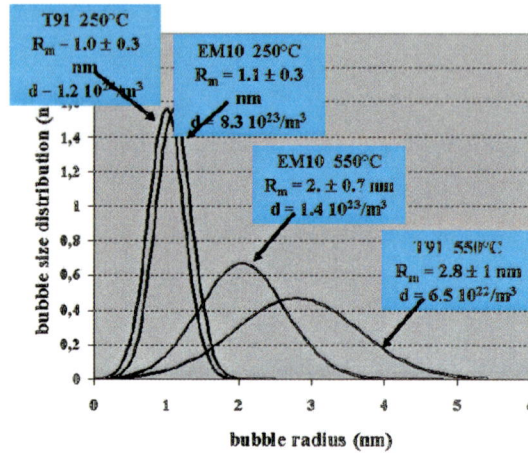

Fig. 4: He-cluster size distribution after implantation at 250 and 500^0C up to 50000 appm He.

TABLE 2: MAIN PARAMETERS OF THE HE BUBBLE DISTRIBUTION OBTAINED BY TEM AND SANS AFTER IMPLANTATION AT 550^0 C IN 9%CR-1MO AND MOD 9%CR-1MO

550^0C	Mean Radius	Standard Deviation	Bubble density
	Rm (nm)	DR (nm)	Nb (m^{-3})
EM10: 9 Cr 1 Mo			
TEM	2.5	0.6	4.2x10^{22}
SANS	2.4	0.4	1.5x10^{23}
T91: Mod 9Cr 1 Mo			
TEM	2.85	0.8	3.0 x0^{22}
SANS	2.8	1	6.5x10^{22}

5.   PREDICTION OF THE LOW TEMPERATURE EMBRITTLEMENT DUE TO POINT DEFECT AND HE ACCUMULATION

The important remaining question is of course: is the characterized microstructure by SANS and TEM able to produce the large hardening due to implantation at 250^0C. ? The simplest model to describe the hardening $\Delta\sigma$ due to a population of obstacles is the Orowan's one:

$$\Delta\sigma = M\alpha Gb(Nd)^{1/2}$$

where N is the bubble density and d the mean radius. M~3 is the Taylor factor due to the fact that we are dealing with a polycrystalline material. α is the obstacle strength, a typical value is 0.3. G=8 x 10^4 MPa is the elastic shear modulus of the matrix. b=0.2 nm is the Bürgers vector. Using N= $1.2x10^{24}$ m^{-3} and d=2 nm resulting from the SANS characterization of the mod 9%Cr-1Mo implanted at 250 ^{0} C (see fig. 4), the calculated hardening is ~705 MPa in good agreement with the tensile data (see fig. 3).

6.   PREDICTION OF RADIATION EFFECTS IN MATERIALS UNDER REACTOR RELEVANT CONDITIONS: THE ROLE OF MULTI-SCALE MODELLING

The behaviour of structural ferritic martensitic steels under fusion relevant conditions i.e. up to high dose and high concentration of He and H will be controlled by the accumulation of point defects and He &H, and, by the phase stability of the ferritic matrix & oxide particles for the ODS steels. It is a complex kinetic process, which involves numerous scale of time and space: from the displacement cascades due to the Primary Knocked-on Atoma (PKA) produced by fast neutrons at the nano-scale for time-duration typical of the pico-second to the evolution of the microstructure controlled by diffusion for time-duration of years for an actual component. Due the progress in the physics of radiation effects and in computer science multi-scale modelling should have an increasing role to understand and separate the various parameters controlling the behaviour of industrial materials under fusion neutron spectrum.

Realistic and reliable multi-scale modelling requires experimental validation at the relevant scale via facilities versatile enough to comply with modelling-designed experiments and with rapid feedback. Ion beam facilities with accelerators in the range of a few MV have got all these characteristics. In the forthcoming years the European Radiation Effect Modelling Programme for martensitic steels under fusion conditions will extensively use this type of facility, especially JANNUS [11], which is being constructed at Orsay and Saclay in France by le Centre de la Recherche Scientifique (CNRS), University of Orsay Paris-Sud and le Commissariat a l'Energie Atomique (CEA).

The validated modelling tools describing the evolution of radiation induced microstructure and the associated dynamics of dislocations should revivify the research & development of radiation resistant materials for fusion. These tools will also be used as reliable guides to (i) inter-correlate the data obtained by the various methods used to simulate the effect of fusion relevant radiation effects, (ii) optimise the testing programme in the 14MeV neutron source International Fusion Materials Irradiation Facility (IFMIF) [12] required to qualify the in-service behaviour reactor structural materials and (iii) extrapolate with improved confidence these 14 MeV neutrons data to the large range of operation conditions of the reactors as it will be necessary in the licensing process.

7.   CONCLUSION

The paper presented the main contributions of Small Angle Neutron Scattering (SANS) provided by Research Reactors to characterize the accumulation of point defects & He and understand the associated hardening under fusion reactor conditions.

The radiation resistance of structural materials is a key issue for the successful development of fusion reactors. The development of realistic multi-scale tools in strong interaction with ion beam irradiation should provide guidance in designing materials with improved resistance, in correlating data from the various neutron spectra and in optimizing the test programme and extrapolating the data from the required 14 MeV neutron source for the qualification of the fusion reactor structural materials.

Up to now the samples characterized by SANS have volumes down to ~10 mm^3 in the best cases. In the strategy foreseen, it should be worth that SANS could have better sensitivity in order to characterize volumes between one or two orders of magnitude lower since (i) the volume irradiated with ion beam facility like JANNUS will be ~0.5 mm^3 , and, (ii) the high activity of the samples irradiated in IFMIF will force to decrease the volumes to be handled.

REFERENCES

[1]   NORAJITRA, P., et. al., . Conceptual Design of the Dual-Coolant Blanket in the Frame of the EU Power Plant Conceptual Study. FZKA Report 6780. November (2002).

[2]   EHRLICH, K., Phil. Trans. Royal Soc. London, A **357** (1999) 595-623.

[3]   SÉRAN, J. L., J. Nucl. Mater. **212-215** (1994) 588-593.

[4]   TAVASSOLI, A. A., J. Nucl. Mater. **329-333** (2004) 257-262.

[5]   JONES, R.H., J. Nucl. Mater. **307-311** (2002) 1057.

[6]   YAMAMOTO,.T, et.al., 7th International Workshop on Spallation Materials Technology (IWSMT-7). to be published in J. Nucl. Mater.

[7]   GELLES, D.S., J. Nucl. Mater. **307-311** (2002) 212-216.

[8]   JUNG, P., HENRY, J., J. Nucl. Mater. **318** (2003) 241-248.

[9]   HENRY, J., MATHON, M.H., JUNG, P., J. Nucl. Mater. **318** (2003) 249-259.

[10]   COPPOLA,R., et.al., J. Nucl. Mater. **329-33** (2004) 1057-1061.

[11]   SERRUYS, Y., Nucl. Instrum. Math. Phys. Res. **B240** (2005) 124-127.

[12]   HEINZEL, V., J. Nucl. Mater. **329-33** (2004) 223-227.

Neutron radiography of advanced nuclear fuels

K.N. CHANDRASEKHARAN, H.S. KAMATH [*]

Radiometallurgy Division, [*] Nuclear Fuels Group, BARC, Mumbai, India

Abstract: Neutron radiography as a specific application in the utilization of India's Research Reactors has been described. The current status of this Nondestructive testing technique in the nuclear energy sector with special emphasis on the testing & characterization of as-fabricated experimental fuel pins, irradiated fuels and structural materials from operating power reactors is presented. Use of this technique for a clear understanding of the performance of the fuel and for a better design of fuel for future reactors have been brought out. Typical results obtained during the course of the neutron radiography activities in the field of nuclear technology in the country are also illustrated.

1. INTRODUCTION

Thermal Neutron Radiography (NR) is one of the specific applications in the utilization of the Research Reactors (RRs) for the Nondestructive Testing (NDT) of nuclear fuels. NR, a relatively new NDT technique, is gaining more importance due to its unique capabilities in material testing and the fact that thermalized neutron beams with adequate intensity are readily available around research reactors. This technique has established itself as a complimentary tool to conventional X ray and gamma ray radiography with diverse applications in nuclear technology, aerospace, ordinance, metallurgy and biology. The unique features of this technique make it possible to inspect the bulk of the material by penetrating high density fuel materials like Uranium & Thorium, to distinguish elements of nearby mass numbers, and to detect small amounts of hydrogenous material formed as a reaction product on metals & alloys. It is possible to detect macro-inhomogenity of PuO_2 in UO_2-PuO_2 & ThO_2-PuO_2 mixed oxide (MOX) fuel matrix. More importantly, this technique also enables volumetric inspection of irradiated fuel and structural materials that are highly radioactive, which may not be possible by other NDT methods due to high radioactivity.

Since the early seventies NR has been pursued in Bhabha Atomic Research Centre (BARC), Trombay, India, using India's first research reactor APSARA, a 400KW swimming pool type reactor, as a neutron source. Thermal neutron radiography has been effectively employed in the Quality Assurance (QA) and the testing of as-fabricated experimental nuclear fuels for complete characterization before irradiation testing so that the data could be compared with that of the post-irradiation results during fuel performance evaluation. Irradiated structural materials, especially Zircaloy pressure tube samples, were evaluated for checking possible Zirconium Hydride formation at the suspected contact points of the Calendria tube in the Pressurized Heavy Water Reactor (PHWR) power reactor. A new facility for the NR of irradiated fuel is presently being set up at the CIRUS research reactor at BARC and a newly commissioned reactor KAMINI at Indira Gandhi Centre for Atomic Research (IGCAR), at Kalpakkam is already being employed for the NR of irradiated fuel pins from the Fast Breeder Test Reactor (FBTR).

This presentation covers applications of the NR of testing and characterization of various types of pre and post irradiated fuel pins, structural materials using the country's various RRs. The further development of NR and the field of nuclear technology also has been briefly described. Typical results obtained during the course of the NR work at BARC and IGCAR in India are also presented.

2. THE RRS USED IN INDIA FOR NR

The APSARA reactor has been in use since 1956 and is the main thermal neutron source for carrying out NR at BARC. In the recently refurbished CIRUS reactor a new facility is being set up for examining irradiated PHWR fuel pins and sections of irradiated Boiling Water reactor (BWR) fuel pins. This facility will also be used for irradiated materials testing and is expected to be ready soon. The recently commissioned reactor KAMINI, a 30 kW swimming pool type reactor, situated at IGCAR Kalpakkam, is presently used for NR of irradiated fuel pins from the operating FBTR.

3.   THE NR TECHNIQUES EMPLOYED

All the three basic techniques viz. Direct, Transfer and Track Etch imaging techniques are well established and regularly employed. Direct technique using Gadolinium screen, transfer technique using Dysprosium converter screen and track etch imaging using Boron coated CN-85 films were employed. Very recently real time NR has been attempted which gave very encouraging results. Neutron Tomography technique is also under development.

4.   NR FACILITY AT APSARA REACTOR, BARC

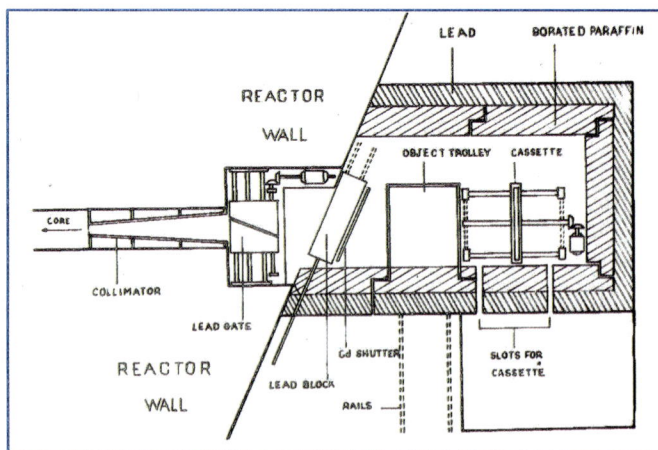

The NR facility at APSARA (Fig.1) has been in use since early seventies [1]. The thermal neutrons from the reactor are collimated by a divergent, cadmium lined aluminum collimator with a length/inner diameter (L/D) ratio of 90. A cadmium shutter facilitates opening and closing of the beam. The specimen can be mounted at about 60 cm from the collimator followed by the cassette containing X ray film and/or converter screen. The whole set up is properly shielded to avoid any radiation exposure to the operating personnel. The important parameters of the facility are presented in Tables 1 (a) and 1 (b), which are schematically shown in Fig.2.

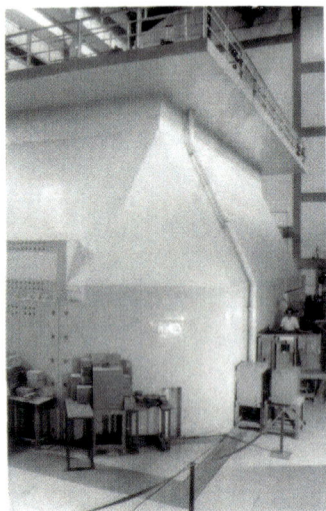

Fig.1 APSARA Reactor   Fig.2 NR Facility at APSARA Reactor (Schematic

TABLE 1(a). APSARA" REACTOR NR FACILITY

Useful beam area	15 cm. Ø
Thermal neutron flux	1×10^6 n/cm^2/sec
Gamma radiation level	4 R/h
Cadmium ratio	6.3
Neutron/Gamma ratio	9×10^5 n/cm^2/mR

TABLE 1(b). A TYPICAL NEUTRON BEAM OF APSARA REACTOR NR FACILITY

Thermal n-content (C)	54 %
Scattered n-content (S)	7 %
Epithermal n-content (E)	3.9 %
Low energy Gamma Ray content	0.5%

5. NR FACILITY AT KAMINI REACTOR, IGCAR, KALPAKKAM

The KAMINI reactor at IGCAR is primarily indented for the Post Irradiation Examination (PIE) of irradiated FBTR fuel pins and subassemblies by NR. This reactor is located below the hot cells meant for irradiated fuel examination (Fig.3). A schematic sketch of the radiographic facility at KAMINI reactor is shown in Fig.4. The facility includes aperture control devices, collimators, beam shutters, film cassette drive mechanism and a radiographic rig for lowering the fuel pins/fuel subassembly in front of the beam tube directly from the hot cells with an indexing mechanism facilitating rotation of the component [2,3]. As the beam hole size is small, for the radiography of long fuel pins a cassette drive mechanism with 10 cassettes arranged in a decagonal fashion and remote indexing has been designed and fabricated. The cassettes are of top loading type and can be loaded or retrieved when the reactor is not in operation. The entire operation was rehearsed with dummy fuel pins. The parameters used for NR at KAMINI reactor is given in the Table-2

TABLE 2. KAMINI REACTOR NR PARAMETERS

Sl. No.	Parameter	Conditions
1.	Technique	Transfer
2.	Screens Used	Dysprosium /100 μ thick
3.	X ray films used	DR-5, AA 400
4.	Reactor Power	15 kW
5.	Exposure	20 minutes
6.	Thermal neutron flux	$10^6 - 10^7$ n/cm^2/sec
7.	Processing	Manual, Standard

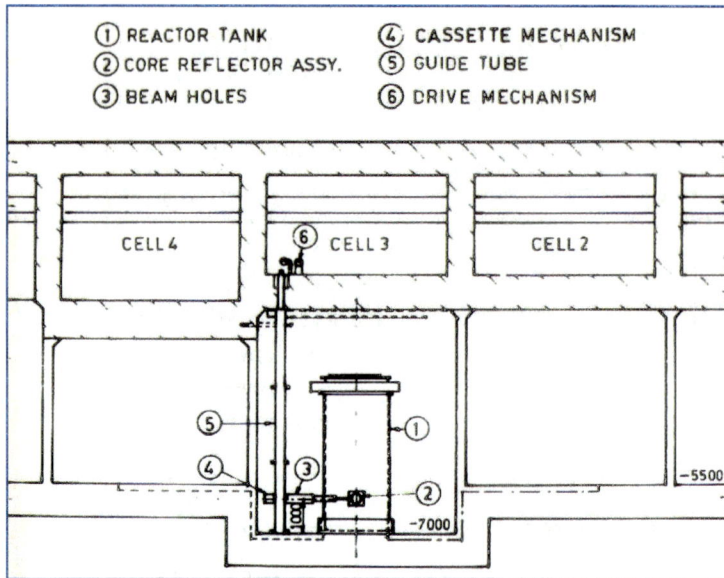

① REACTOR TANK   ④ CASSETTE MECHANISM
② CORE REFLECTOR ASSY.   ⑤ GUIDE TUBE
③ BEAM HOLES   ⑥ DRIVE MECHANISM

CELL 4   CELL 3   CELL 2

−5500

−7000

Fig.3 KAMINI Reactor housed below the Hot Cell

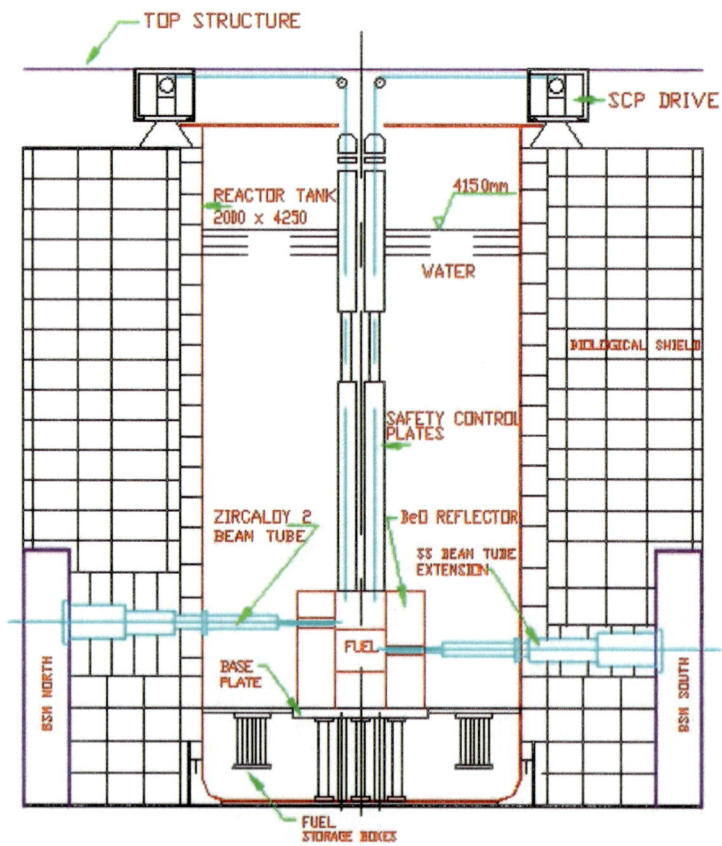

TOP STRUCTURE

SCP DRIVE

REACTOR TANK
2000 x 4250

4150mm

WATER

BIOLOGICAL SHIELD

SAFETY CONTROL
PLATES

ZIRCALOY_2
BEAM TUBE

BeO REFLECTOR

SS BEAM TUBE
EXTENSION

FUEL

BSS NORTH

BASE
PLATE

BSS SOUTH

FUEL
STORAGE BOXES

Fig.4 KAMINI Reactor NR facility (Schematic)

6. OTHER NR FACILITIES IN INDIA

Defence laboratory, Jodhpur, operates a radioisotope based (Cf [252]) NR facility whereas SHAR Centre-ISRO, Shriharikota, has an accelerator based NR facility. Both the facilities are in use for defence and space applications.

7. NR OF NUCLEAR FUELS IN BARC

7.1 As-fabricated nuclear fuel elements

A typical experimental nuclear fuel pin essentially consists of a thin walled cladding tube filled with UO_2, UO_2-PuO_2 or PuO_2-ThO_2 MOX fuel pellets and hermetically sealed at both ends by welding of end plugs (Fig. 5). These experimental fuel pins are used for irradiation testing in research reactors to study the fuel performance. Enriched uranium-aluminum alloy fuel sandwiched between aluminum-clad plates and roll-bonded into fuel plates are used as core in small research reactors (like KAMINI) employed for physics research and for applications like neutron radiography.

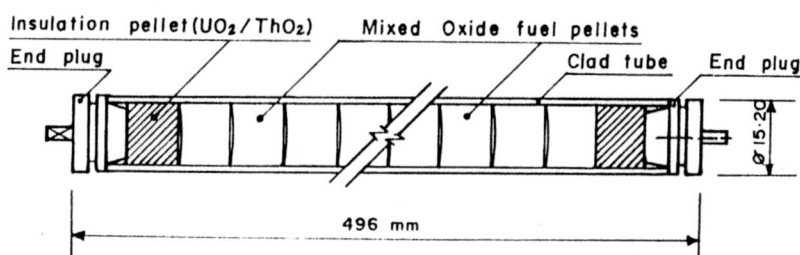

Fig.5 Experimental fuel pin (Schematic)

7.2 Observations

Interesting results have been obtained in NR work at APSARA on experimental fuel elements [4-9]. NR was carried out on various types of experimental fuel elements conforming to PHWR type, BWR type, Fast reactor (FBTR type) and alloy plate fuel elements for KAMINI reactor. The experiments were conducted on the sealed fuel elements to obtain information such as:

a) Integrity of the pellets in as-fabricated fuel pin

b) Limit of detection of central voids in fuel pellets

c) Water ingress into the fuel pin

d) Compositional variation of fissile fuel loading

e) Identifying solid and annular fuel pellets in case of a mix up

f) Detection of PuO_2 agglomerates in a PuO_2-UO_2 mixed oxide fuel pellet

g) To check the homogeneity of distribution of fissile isotope in Al-U [233] alloy fuel plates

Direct technique using 25μ Gd screen, transfer technique using 100μ Dy converter screen and track-etch technique using Kodak CN85 type B coated cellulose nitrate films were carried out for generating these information. Typical exposure parameters used for MOX fuel pins are presented in Table 3

TABLE 3. APSARA REACTOR NR EXPOSURE DETAILS

(1) Direct Technique:

Screen	Gd / 25 μ thick
X ray film	Agfa structurix D4
Neutron fluence	3×10^8 n/cm^2
Exposure time	5 min
Developing	5 min at 20 ^{0}C

(2) Transfer Technique:

Screen	Dy / 100 μ thick
X ray film	Agfa structurix D4
Neutron fluence	4.2×10^9 n/cm^2
Exposure time	70 min
Transfer time	70 min
Developing	5 min at 20 ^{0}C

(3) Track-etch technique:

Film	Kodak CN 85 type B (double coated)
Neutron fluence	1.5×10^9 n/cm^2
Exposure time	25 min
Etching solution	NaOH (6N) at 55 ^{0}C
Etching time	40 min

In PHWR type reactors, the fuel pin contains natural uranium dioxide in the pellet form (diameter 14.30mm nominal) encased in a Zircaloy cladding tube and sealed at both ends by welding of end plugs. NR was carried out to check the integrity of the as-fabricated fuel pins as conventional X ray radiography fails to penetrate such high density fuel pellets. In order to reduce the center temperature of the fuel pellet during irradiation and thereby reduce the pellet clad mechanical interaction, annular pellets (pellets with a central void) or pellets with depleted UO_2 central core are being considered in BWR fuel pins. The deep penetration of neutrons in dense and high Z material is utilized to identify these pellets. Fig.6 presents an experimental fuel pin containing annular and solid pellets inside a fuel pin. Water ingress inside a fuel pin because of a breach in the clad has been simulated and a typical radiograph of the same is presented in Fig.7.

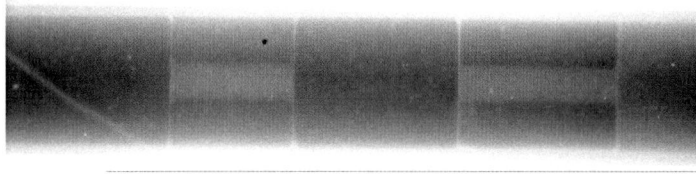

Fig.6 NR of solid and annular fuel pellets inside a fuel pin

Water ingress

Fig.7 Water ingress inside a PHWR type fuel pin (Simulated)

The unique capability of NR in discriminating nearby elements has been utilized to identify UO_2 and PuO_2 -UO_2 mixed oxide fuel pellets inside sealed experimental fuel pins. Fig.8 presents the radiograph of a set of MOX fuel pins containing UO_2 and UO_2 - PuO_2 MOX pellets. The damaged UO_2 insulation pellets at the end of the fuel stack can be clearly seen in the radiograph (Track-etch radiograph). During mixing of PuO_2 with UO_2 for making MOX pellets, there is a possibility of PuO_2 agglomerates being present as inclusions in the MOX matrix. Homogeneity of mixing could be checked by this technique. NR reveals PuO_2 agglomerates in MOX matrix as depicted in Fig.9. The limit of detection of PuO_2 agglomerate in low Pu-enriched MOX fuel was found to be of the order of 250µ. Figure 10 presents the radiograph of an experimental MOX fuel pin containing UO_2 (dark) and UO_2-Gd_2O_3 (white) fuel pellets.

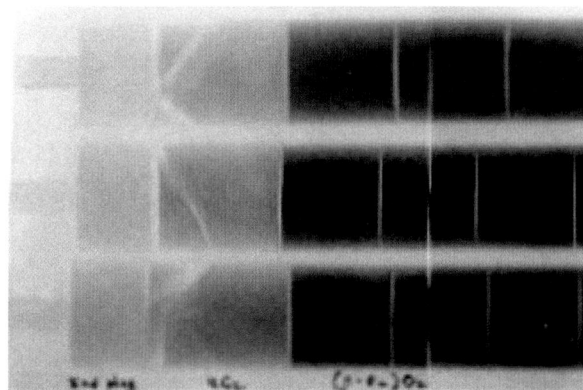

Fig.8 Damaged UO_2 pellets inside a set of Advanced PHWR type fuel pins (MOX)

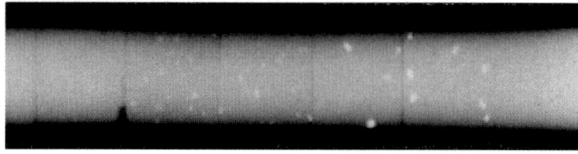

Fig. 9 PuO₂ agglomerates as inclusions inside MOX fuel pellets

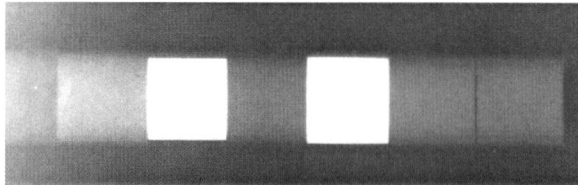

Fig. 10 NR of an experimental fuel pin containing UO₂ (dark) and UO₂-Gd₂O₃ (white) fuel pellets.

In fast reactors, the fuel pellets, which are small in diameter, are encapsulated in SS 316 cladding tubes along with other components. During loading of fuel pellets inside the thin walled clad tubes and subsequent operations like end plug welding, decontamination, wire wrapping etc., the pellets may get chipped or cracked and the chips may get trapped in the pellet/clad gap or pellet/pellet gap. NR was carried out on as-fabricated BTR type experimental fuel pins to check the fuel pellet integrity. Figure 11 presents a set of FBTR type fuel pins showing damaged fuel pellets (Track-etch radiograph).

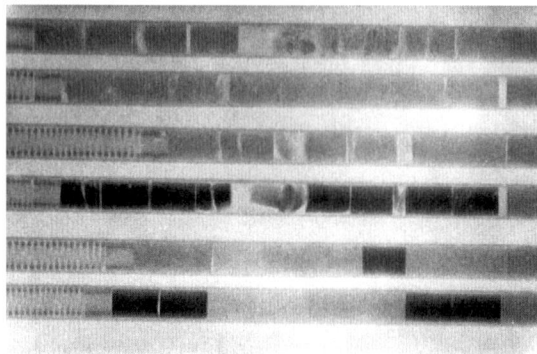

Fig. 11 Damaged fuel pellets inside FBTR type fuel pins

Distribution of fissile isotope in an Al-U^{233} alloy fuel plate was also evaluated using NR. As the fissile fuel content in this type of alloy-plate is less and highly absorbing to neutrons, the homogeneity variation in the alloy plate results in optical density variation in the radiograph. NR of Al- U^{233} alloy fuel plate showed uniform distribution of U^{233} in aluminum matrix.

The use of a neutron opaque material to improve the contrast of the defects / profile of the inserts and shapes of heavy castings of high-density material was also explored. This provided a contrast improved primary image, which is amenable for further processing (Fig.12(a) & Fig.12(b).

Coating

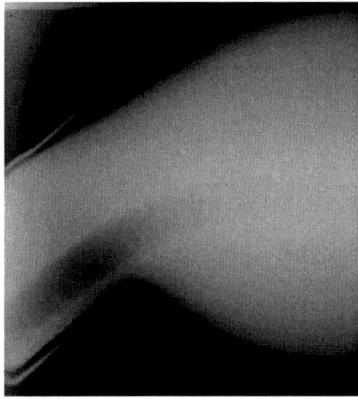

Fig.12(a) Normal radiograph
of a casting with a tube insert

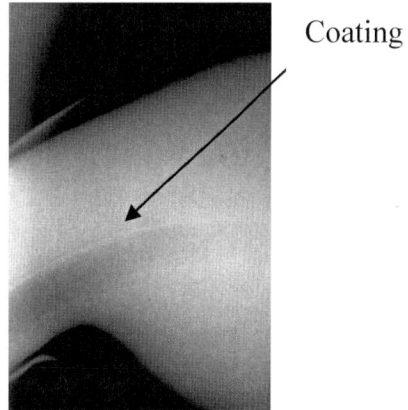

Fig.12(b) Radiograph with neutron
opaque material coating inside the tube

8. NR OF IRRADIATED FUELS AND STRUCTURAL MATERIALS AT IGCAR AND BARC

8.1 Irradiated fuel pins

One of the most important applications of NR in the nuclear field is in the post irradiation examination of the irradiated nuclear fuels. As the associated radiation level in the irradiated fuel is very high, conventional radiography methods cannot be used for checking this fuel. In the nuclear fuel development programme, pre-irradiated, interim and post-irradiated fuel radiographs provide valuable information about the fuel performance under various conditions. Information like fuel cracking, swelling, melting, void formation and fissile redistribution etc. can be obtained qualitatively and quantitative measurements can be extracted from the radiographs. The feed back of this helps in a better design of the fuel.

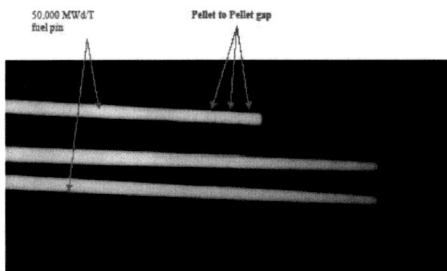

Fig.13(a) Neutron Radiograph of
Irradiated FBTR fuel pins

Fig. 13(b) Neutron Radiograph of
Irradiated FBTR fuel pins
(Detail showing fuel pellet-pellet gap)

The facility for NR of irradiated fuel pins is already functional at IGCAR. Radiography of irradiated FBTR fuel pins is regularly being carried out. Irradiated fuel pins withdrawn after 50 GWD/T of burn-up has been completed. Typical radiographs of the irradiated FBTR fuel pins are presented in Fig.13(a) and the pellet-pellet gap as seen in the radiograph is presented in Fig.13(b). Presently, efforts are being made for the radiography of fuel pins that have undergone 100 GWD/T burn-up.

The other facility in BARC for the NR of irradiated fuel pins from power reactors (PHWR & BWR) is nearing completion. The irradiated fuel pins and cut sections selected during PIE of these fuels inside a hot-cell shall be transported to the NR site of the CIRUS reactor in a fuel transport cask [10]. The necessary shielding and operation facility is incorporated in three

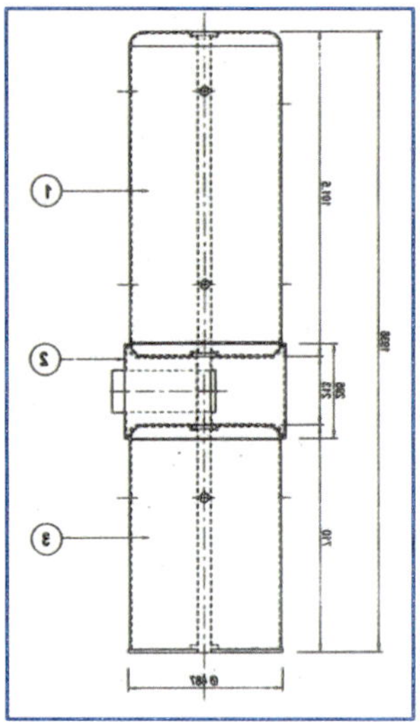

Fig.14 Capsule used for the NR of irradiated fuel pins (Assembled in three parts)

parts (Fig.14). The lower cask shall be permanently positioned in the NR site. The upper cask is used for fuel transportation from the hot-cell to the NR site and back. This part of the cask will be positioned above the stationary part and after mating perfectly; the irradiated fuel pin will be lowered to the middle portion for radiography. Necessary safety precautions as per international regulations are strictly adhered to. Presently transfer technique using Dy screen is being planned. In order to cover the long fuel pins a cassette magazine is designed and the fuel pin will be lowered to the required position each time so that the entire length is scanned. The same facility is proposed to be used for selected irradiated pressure tube samples for detection and characterization of hydride blisters by NR.

8.2   Zircaloy pressure tube (PT)

Zirconium hydride blister formation on Zircaloy PT, used in large numbers in PHWR power reactors, is of a safety concern. The contact of PT with the Calendria tube (CT) results in the hydrogen migrating to the cold contact locations and subsequent formation of hydrogen blisters. The blisters would eventually grow and crack. The formation of a series of such cracked blisters might lead to a rupture of PT by the delayed hydrogen cracking mechanism. The unique feature of NR in detecting low 'Z' materials in high 'Z' metal matrix is taken advantage in the detection of hydrides.

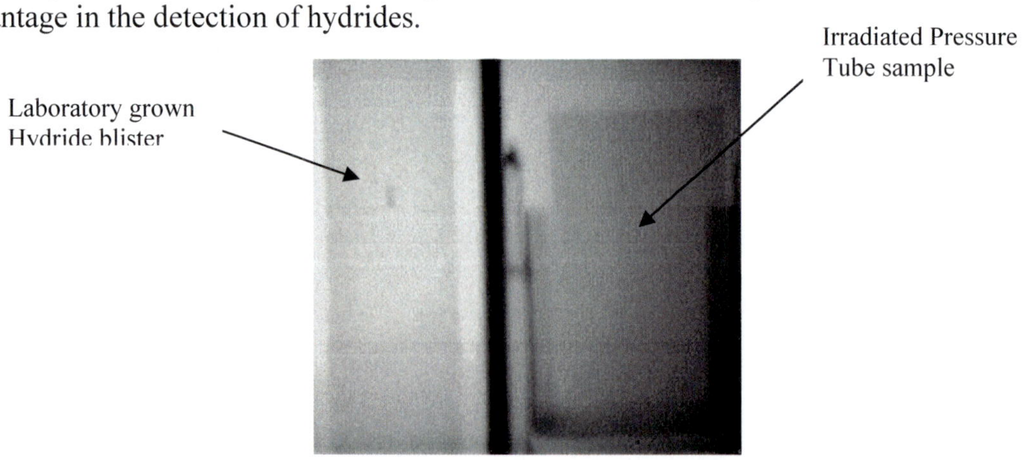

Fig.15 NR of irradiated Zircaloy Pressure Tube sample along with the laboratory grown Zirconium Hydride blister

NR was employed for the detection and characterization of laboratory-grown blisters and compared with that of irradiated pressure tube samples [11,12]. Micro-densitometry scanning of the radiographs provided sizing of the blisters. Zircaloy PT samples selected during the post-irradiation examinations of the tube inside the hot-cell was radiographed along with the laboratory grown blisters for comparison (Fig.15). A comparison of this data with that of ultrasonic testing results has shown a very god correlation. Further work in this area is under way.

9. QUANTITATIVE ANALYSIS OF RADIOGRAPHS (QNDE)

9.1 Dimension Measurement from Radiographs

The most important dimensions measured from radiographs are fuel pellet diameter & length, pellet-clad gap, pellet-pellet gap, dishing into the pellet and fuel central channel diameter. An accurate dimension measurement from the radiograph was carried out using a scanning microdensitometer. A specially fabricated calibration fuel pin containing UO_2 sintered fuel pellets (14.20mm nominal diameter) with known central channels of varying diameter was used for the evaluation of dimension estimate from the radiographs [13]. The true dimensions were measured with an accuracy of $\pm$ 0.001mm. Typical microdensitometer scan of a solid pellet and those pellets with central channels of known dimensions are presented in. Fig. 16. The smallest central channel diameter detected was of 1.60mm. It was observed that the deviation increases as the central channel diameter to be measured decreases. Track-etch method gave diameter estimate close to the true value in comparison to transfer technique. NR was carried out using transfer and track-etch techniques. Image magnification factor (M = 1.003344) arising from the geometry of the NR set up was considered for calculating the deviation of the measured values from the true values. For the similar configuration of irradiated fuel pellet, by applying the correction factor the true dimensions could be found out.

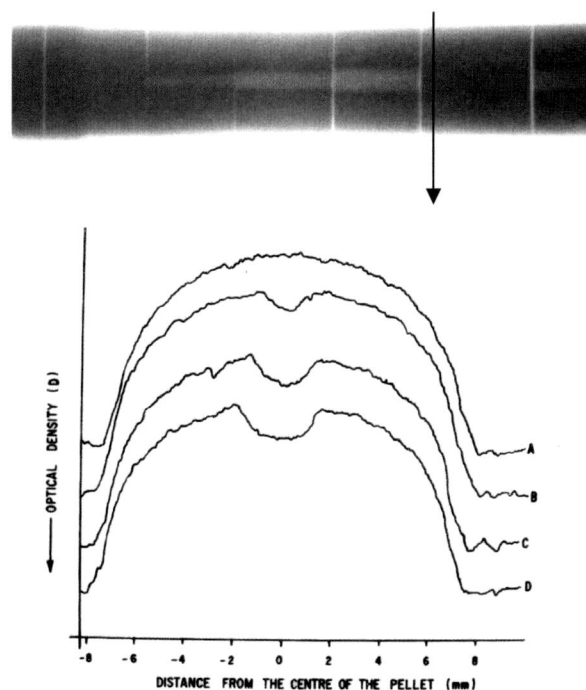

Fig.16 NR of the calibration fuel pin containing fuel pellets with varying central channels and respective microdensitometer scans across the pellets A – Solid pellet; B,C,D – Pellets with central channels

Optical density data corresponding to different thickness of the pellet starting from the center towards the periphery were generated by scanning microdensitometry and thermal neutron interaction probability was correlated to the optical density at points corresponding to different thickness segment of the fuel pellet [14]. The density data was used to postulate a simple model for neutron radiography of fuel pins to predict the optical density in neutron radiographs of fuel pins with pellets of different compositions and diameters [15].

9.2 Digital image processing

Digital image processing has been employed for delineating PuO_2 inclusions in a UO_2-PuO_2 MOX fuel pellet. Fig.17 presents an edge-enhanced radiograph showing PuO_2 particles in MOX pellets. Microdensitometer traces of MOX pellets having various percentages of PuO_2 in UO_2 are presented in Fig.18. A good correlation between the optical density of the neutron radiograph over the fuel pellets and PuO_2 enrichment was established. Using this, PuO_2 enrichment in an unknown MOX pellet inside the fuel pin could be determined.

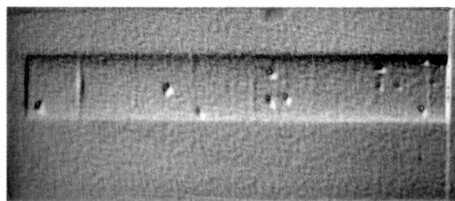

Fig. 17 Image enhanced NR showing PuO_2 agglomerates in MOX fuel pellets

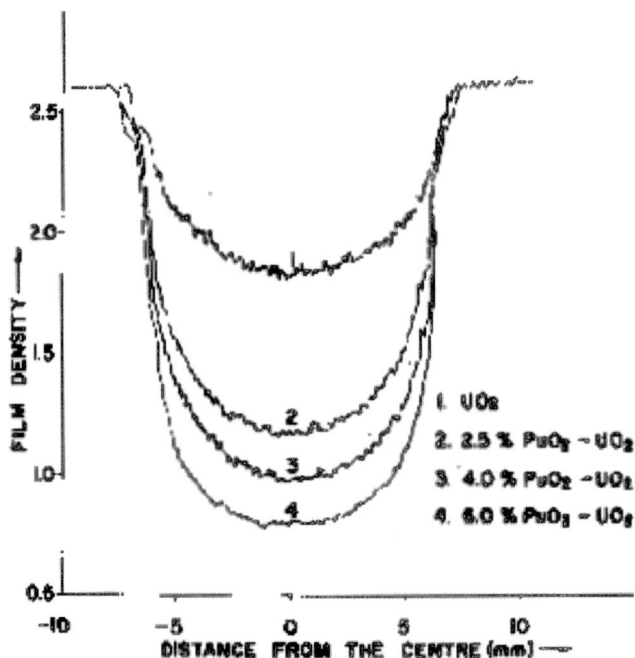

Fig. 18 Microdensitometer traces of MOX fuel pellets havingvarious percentages of PuO_2 in UO_2-PuO_2 MOX pellets

9.3 Evaluation of the Resolution Capability of APSARA NR Facility.

NR facility at APSARA reactor was evaluated using a resolution parameter lambda (λ) proposed by A.A. Hams [16]. The knowledge of such a single parameter λ could be used as a performance capability index of a specific NR facility. Edge Spread Function of a Cadmium knife-edge image obtained by scanning microdensitometer was used for the evaluation of λ and for the measurement of the total Unsharpness (U_T) [17]. The λ values and the U_T for various screen-film combinations of the facility were evaluated. The effect of change in L/D ratio on the λ value was also evaluated. It was observed that the λ value decreases with L/D ratio. The empirical method of determining the magnitude of un-sharpness by Klasens method [18] was employed to obtain the U_T. The optical density scan across the Cd knife-edge was used for evaluating the U_T.

10. CONCLUSION

NR as one of the specific applications in the utilization of RRs for materials testing and characterization of nuclear fuels is described. Thermal NR using India's research reactors ARSARA, CIRUS at BARC and KAMINI at IGCAR, Kalpakkam has been explained emphasizing applications in the nuclear energy sector. NR of as fabricated and irradiated nuclear fuel pins and structural materials are explained. New developments in the field of NR like Neutron Tomography at BARC, control rod NR at IGCAR and QNDE using microdensitometer scanning and digital image processing are briefly mentioned. Typical results obtained during the course of NR work are also illustrated.

11. ACKNOWLEDGEMENTS

The authors wish to acknowledge sincere thanks to their colleagues in Nuclear Fuels Group, BARC, particularly to the staff of Inspection & Quality Control Section, Radiometallurgy Division & Post Irradiation Examination Division for carrying out the work. The contributions provided by the Materials & Metallurgy Group of IGCAR, particularly to the staff of Post Irradiation Examination & Robotics Division, are acknowledged. The support rendered by the reactor superintendents of APSARA and KAMINI reactors is also acknowledged.

REFERENCES

[1] DANDE, Y.D.,Neutron Radiography; BARC-768 (1974)

[2] KASIVISWANATHAN, K.V., VENKATRAMAN, B., BALDEV RAJ, Neutron Radiographic Facilities Available with KAMINI, Proc. of 6th World Conference on NeutronRadiography, Edited by Shigenori Fujine, Hisao Kobayashi and Keiji Kanda, Gordon and Breach Science Publishers, Japan, 1999, pp. 117-120.

[3] VENKATARAMAN, B., et.al., Radiographic Techniques for Post Irradiation Characterization of Fast Breeder Test Reactor Fuel Pins, Presented at the Intl. Conf. on Quality Control of Nuclear Fuels, (CQCNF), Dec.10-12, 2002, Hyderabad, India.

[4] GHOSH, J.K., PANAKKAL, J.P.,ROY,P.R., Monitoring plutonium enrichment in mixed oxide fuel pellets inside sealed nuclear fuel pins by Neutron Radiography; NDT International Vol.**16** No.5 (1983) 275-276,.

[5] GHOSH, J.K., PANAKKAL, J.P.,ROY, P.R., Characterisation of uranium plutonium mixed oxide nuclear fuel pins using neutron radiography; British J. of NDT, **27** (1985) 232-233.

[6] GHOSH, J.K., PANAKKAL, J.P., ROY, P.R.,Quantitative Non Destructive Evaluation of sealed Nuclear Fuel Elements; International Advances in NDT (USA), Vol:12 pp 53-70; (.McGONNAGLE, W.J., Ed), Gordon and Breach (1986).

[7] DANDE, Y.D., et.al.,NeutronRadiography at APSARA reactor; Indian J. of Pure & Applied Physics; vol.**29** (1991) 721-731.

[8] CHANDRASEKHARAN, K.N.,PATIL, B.P., GHOSH, J.K., Neutron Radiography for characterisation and quality assurance of nuclear fuel pins; J. of Nondestructive evaluation; **15**, no.1 (1995) 1-9.

[9] GHOSH, J.K., et.al., Neutron Radiography of nuclear fuel pins - An album; BARC-1997/E/019.

[10] SAHOO, K.C., GANGOTRA, S., BALAN, T.S., SIVARAMAKRISHNAN K.S., "Neutron Radiography Facility for Irradiated Fuels", paper presented at IAEA CRP

meeting on "Examination and Documentation Methodology of Water Reactor Fuels" during February 21-24, 1989, Vienna.

[11]   GANGOTRA, S.,. et.al., Detection of Hydride Blisters in Irradiated Zircaloy Pressure Tube by Neutron Radiography, 7th European Conference on Non Destructive Testing, 26 – 29 May 1998, Copenhagen, Denmark.

[12]   GANGOTRA, S., OUSEPH, P.M., SHAIKH, A.M., SAHOO, K.C.,Neutron Radiography of Contact Location of Irradiated Zircaloy-2 Pressure Tube from RAPS-II, Proceedings of the 14th World Conference on Non Destructive Testing, New Delhi, Dec 8-13, 1996, Vol.**3**, p1453-1456.

[13]   CHANDRASEKHARAN, K.N., PATIL, B.P., GHOSH, J.K., Quantitative evaluation of structural features from nuclear fuel pin neutron radiographs; Neutron Radiography (4); Proceedings of the 4th World Conf. on Neutron Radiography, San Francisco, 10-16 May.1992, pp 249-256, (BARTON, J. P., Ed.), Gordon & Breach Science Pub. USA, 1994.

[14]   PANAKKAL, J.P., GHOSH, J.K., ROY; P.R., Analysis of optical density data generated from neutron radiographs of uranium-plutonium mixed oxide sealed nuclear fuel pins; Nuclear instruments and methods in physics research: **B14** (1986) 310-313.

[15]   J.P. PANAKKAL, J.K.GHOSH; A simple model for Neutron Radiography of Uranium - Plutonium mixed oxide fuel pins,, J. of Nuclear Materials, **153**. No:1-3 (1988) 82-85 .

[16]   HARMS A. A., WYMAN D. R. Mathematics and Physics of Neutron radiography; D. Ridel Pub. Co. Dordrecht, Holland (1986).

[17]   CHANDRASEKHARAN, K.N., PATIL, B.P., GHOSH, J.K.,Evaluation of the resolution capability of APSARA Neutron radiography facility at BARC; Neutron Radiography (5); Proceedings of the 5th World Conf. on Neutron Radiography, Berlin,17-20 ,June.1996.

[18.]   KLASENS. H.A; "Measurement and calculation of Unsharpness combinations in X ray photography"; Philips Research Reports, Vol.**1**,no.4 (1946) pp241-149.

Utilization of 14 MW TRIGA research reactor integrated in a structure for materials and nuclear fuel characterization and development

C. PAUNOIU, M. CIOCANESCU, M. PÂRVAN, M. MINCU, O. UŢĂ, S. IONESCU

Romanian Authority for Nuclear Activities – RAAN, Institute for Nuclear Research, ICN Piteşti, Romania

Abstract: The Institute for Nuclear Research (ICN) of Piteşti has a set of nuclear facilities consisting of TRIGA 14 MW(th) materials testing reactor and LEPI (Romanian acronym for post-irradiation examination laboratory) which enable to investigate the behaviour of the nuclear fuel and materials under various irradiation conditions. The LEPI is an alpha-gamma hot cell facility able to manipulate and examine radioactive materials having an activity up to 10^6 Ci ($E_{average} \leq 1$ MeV) and a high content of transuranium elements (Pu, Am, Cm). In order to obtain relevant informations on CANDU nuclear fuel performance, a significant number of fuel elements manufactured by ICN has been tested to different power histories in the TRIGA 14 MW(th) reactor. Most important tests have been performed in conditions of power ramping, overpower and accident. After testing, the fuel elements have been examined in the hot cells at LEPI using various post-irradiation examination techniques. These techniques include both non-destructive methods (visual inspection and photography, eddy current testing, profilometry, gamma scanning) and destructive methods (fission gas release and analysis, matallography, ceramography, burnup determination by mass spectrometry, mechanical testings). The data obtained from post-irradiation examinations are used on one hand to confirm the integrity, safety and performance of the irradiated fuel and on the other hand for further progress in CANDU fuel development.

1. INTRODUCTION

The development of the nuclear energy in our country by commissioning of new units at Cernavoda Nuclear Power Plant (NPP) requires the nuclear fuel safety strengthening and implicitly the development and improvement of the techniques to investigate the processes which take place in the fuel, in order to obtain relevant informations concerning:

- Maneuvre regimes or power rampings;
- Nuclear fuel behaviour in accident conditions generated by reactivity and loss of coolant;
- Fuel element failure kinetics and fission product release.

These informations are needed to evaluate the performance of nuclear fuel and materials in NPP in order to check the concordance with safety criteria.

Safety criterion used in power ramping conditions is the "Pellet-Clad Mechanical Interaction" (PCMI) which is related to the stress on the cladding produced by the UO_2 pellet expansion in a short period of time. This situation can result in an interaction and if the stress is high enough and the cladding ductility low enough the „ramp" situation can lead to a clad failure.

In the CANDU reactor this type of event is of particular importance because this type of reactor uses the „on-power" re-loading and therefore the CANDU fuel endures fairly severe ramps all the time.

In order to check and improve the quality of the Romanian CANDU fuel, the power ramping tests have been performed on experimental fuel elements in the TRIGA SSR (Steady State Reactor) of SCN Piteşti and their behaviour has been analysed by post-irradiation examination (PIE) in the hot cells.

The accident conditions which can lead to fuel melting are the loss of coolant or an excessive power in the fuel. If these accidents happen, the coolability of reactor core and its geometry must be preserved, thus avoiding the fuel dispersion and major consequences on the environment. The accidents studied to prevent these possibilities are:

- Loss of Coolant Accident (LOCA)
- Reactivity Insertion Accident (RIA)

In order to investigate the Romanian CANDU fuel behaviour in the accident conditions, simulated LOCA and RIA tests have been performed on experimental fuel elements in the TRIGA ACPR (Annular Core Pulsed Reactor) reactor of SCN Piteşti.

2. POST-IRRADIATION EXAMINATION TECHNIQUES FOR PERFORMANCE EVALUATION OF IRRADIATED NUCLEAR FUEL

The nuclear fuel used in the CANDU-6 reactor at Cernavoda NPP consists of 37 elements assembled in the shape of bundle by means of end plates. This fuel bundle have the length of 495 mm, the diameter of 103 mm and the weight of 24 kg (Fig.1).

Fig.1 CANDU fuel bundle

CANDU fuel element contains sintered cylindrical pellets of uranium dioxide (UO_2) stacked within a Zircaloy-4 cladding tube sealed at both ends by end plug. Its length is of 492 mm and its diameter is of 13.08 mm.

In the nuclear reactor the fuel elements endure dimensional and structural changes as well as sheath oxidation, hydriding and corrosion. These changes can lead to defects and even to the loss of integrity.

The nuclear fuel performance is determined by the following elements:

- Surface condition (corrosion product deposition, corrosion etc.);
- Sheath integrity;
- Dimensional changes;
- Burnup;
- Fission product activity distribution in the fuel column;
- Pressure and volume of fission gas;
- Structural changes of fuel and sheath;
- Oxidation and hydriding of sheath;
- Fuel isotopic composition;
- Sheath mechanical properties.

2.1 Non-destructive post-irradiation examination techniques

The non-destructive PIE of the nuclear fuel into the LEPI hot cells includes:

- Visual inspection and photography;
- Profilometry (diameter, bow and ovality) and length measurement;
- Gamma scanning and Tomography;
- Eddy current testing.

The main examination equipments consist of three fuel rod-positioning machine: one for visual inspection, one for profilometry and eddy current testing and one for gamma scanning. These machines are equipped with step-by-step motors to carry out the vertical movement and rotation of the fuel element and are remotely controlled by means of control desks which display digitally the fuel element position.

2.1.1 Visual Inspection and Photography

A periscope set in the hot cell shielding wall having 4 magnifications: x2, x4, x6 and x12 is used to examine visually the irradiated fuel elements. A digital camera is used to take photos by the periscope eyepiece.

The purpose of visual examination is to observe the macroscopic changes of the fuel element surface condition such as failure, corrosion product deposition, corrosion, swelling etc. due to both manufacturing conditions and irradiation conditions (Fig.2 and Fig.3). Figure 3 shows that the fuel element has been cracked due to severe reactivity insertion test endured in the TRIGA pulse reactor.

Fig.2 CANDU fuel element irradiated in the TRIGA reactor in a power ramping test (x0,5)

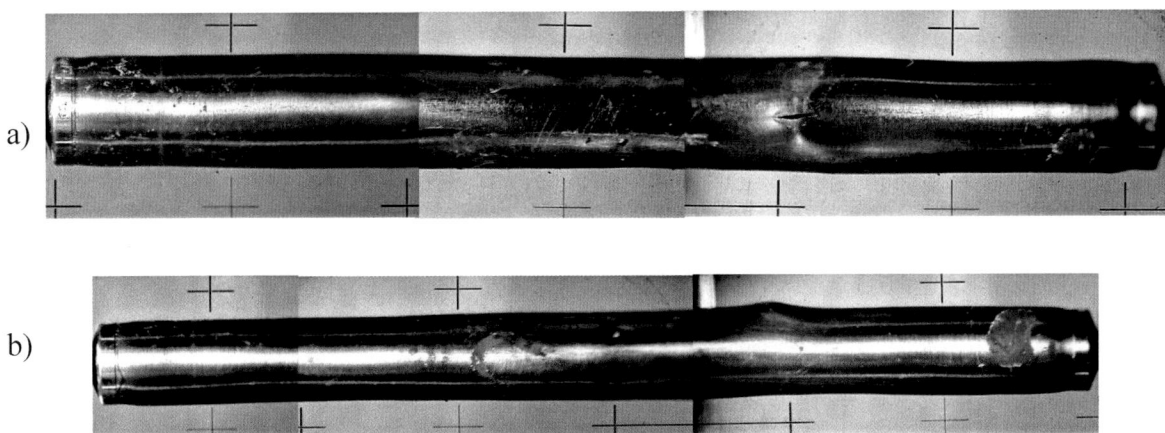

Fig.3 CANDU fuel element irradiated in the TRIGA pulse reactor in a RIA test a) rotation 0^0 (crack direction); b) rotation 90^0

2.1.2 Profilometry

The dimensional changes of the fuel elements during irradiation are an important parameter to evaluate the CANDU nuclear fuel behaviour and performance in operation. They are diametral swelling, ridging, elongation, bending and ovalization of the fuel element and are the result of the swelling of Zircaloy-4 sheath, swelling of fuel and fuel-cladding interactions induced by nuclear radiations.

Profilometry is a step-by-step measurement of the diameter along the fuel element at regular intervals (usually 1 mm) using a remotely operable profilometer piloted by computer. Two opposed inductive transducers being differential transformers having a mobile core are used. The diameters are measured by vertical movement of the fuel element in steps with the plunger cores of the two opposed transducers touching it. Thus, a diameter profile of fuel element is obtained. The accuracy of the measurement is ±5 μm and is obtained by linearization of transducers.

Figure 4 shows the profilometry results obtained from an experimental CANDU fuel element both before and after irradiation. This fuel element (shown in Fig.2) has been irradiated in the TRIGA reactor in a power ramping test. Sheath ridging is observed due to pellet end swelling, a typical phenomenon of CANDU fuel. Also, it can observe the sheath swelling as well as the fuel element bending.

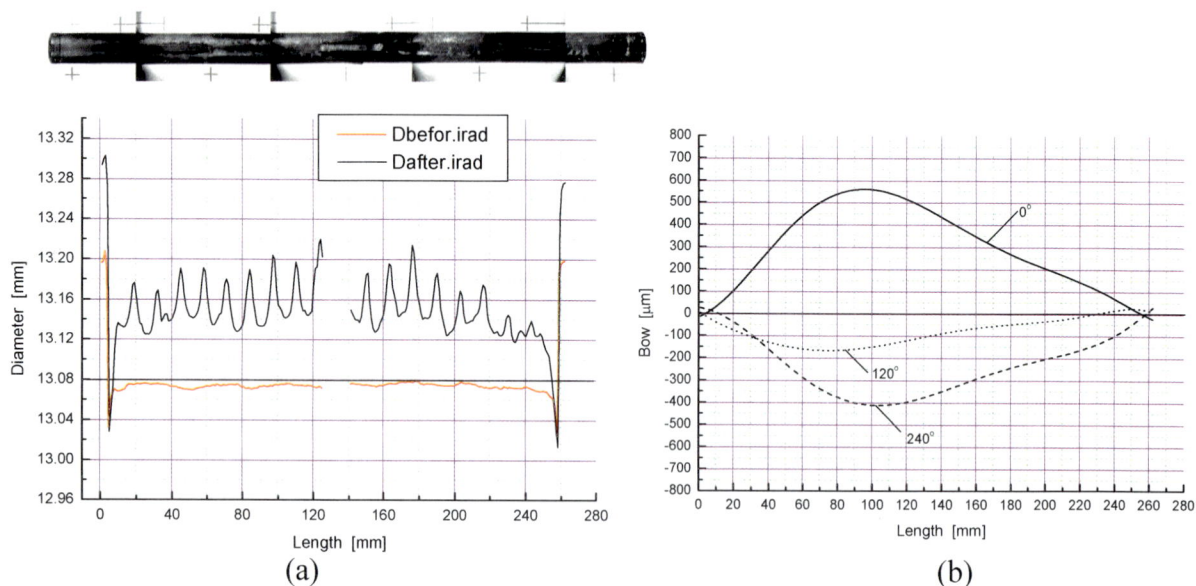

Fig.4 Average diametral profile before and after irradiation (a) and bending profile after irradiation (b) of a candu fuel element tested in a power ramping in the TRIGA reactor

- Average diameter before irradiation: 13,077 mm   Bow: 0,573 mm, at
                                                                     position 98 mm, on direction 0^0
- Average diameter after irradiation: 13,150 mm
- Maximum diametral swelling: 0,146 mm
                              at position: 124 mm
- Average diameter at top of ridges: 13,191 mm
- Average diameter at bottom of ridges: 13,131 mm
- Average height of the ridges: 30 μm

Figure 5 shows the profilometry results obtained from the experimental CANDU fuel element shown in Fig.3 which has been irradiated in the TRIGA pulse reactor in a simulated RIA test. Prominent sheath swelling is observed due to fuel swelling and a violent fission gas release which have determined sheath cracking. The sheath ovalization profile has been obtained by circumferential diameter measurements.

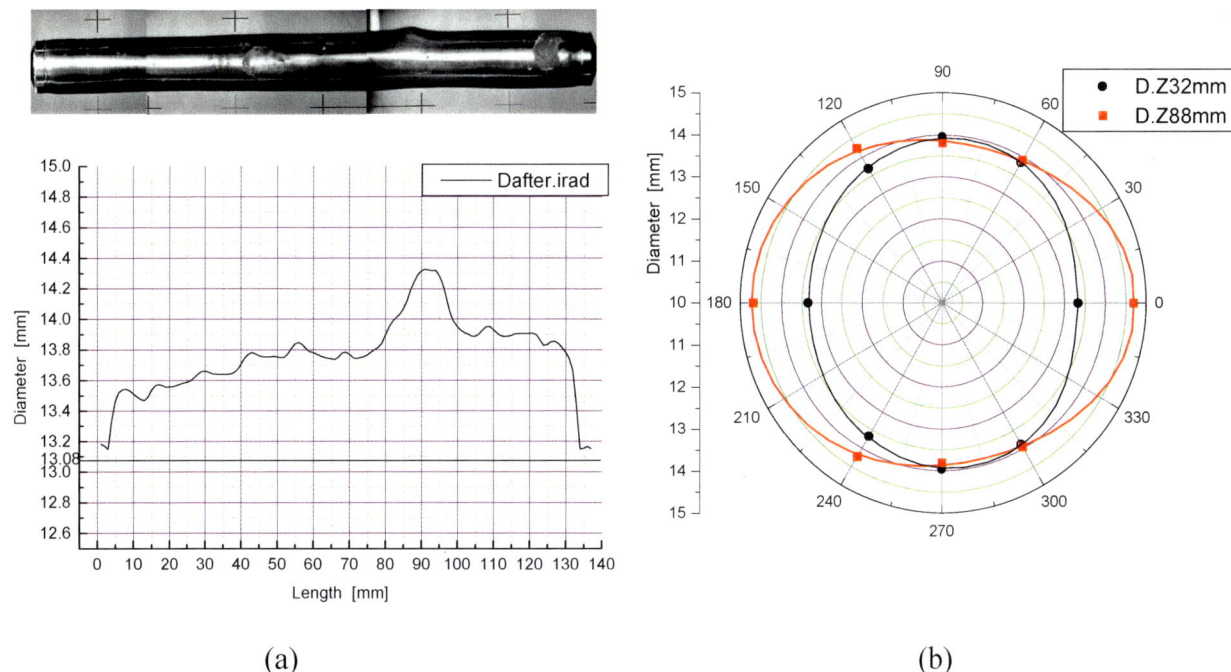

(a) (b)

Fig.5 Average diametral profile (a) and ovalization profile (b) of a candu fuel element tested in a simulated ria in the triga pulse reactor

- Maximum diameter:	14,326 mm		Cota 88 mm:
at position:	91 mm	- Maximum diameter:	14,685 mm
- Average diameter:	13,768 mm	- Average diameter:	14,223 mm
- Maximum diametral swelling:	1,246 mm (9,5%)	Circumferential elongaion:	3,6 mm(8,7%)

2.1.3 Gamma Scanning and Tomography

The gamma scanning technique offers a rapid method to determine the axial and radial distribution of the fission products (FPs) activity in the fuel element, the migration of volatile FPs, the geometrical characteristics of the fuel column and the burn-up. The fuel burn-up can be determined with an accuracy of ±10 %. The acquisition system is composed of a collimator with a variable slit which can be on the horizontal or vertical position, a PGT intrinsic Ge detector and a multichannel analyser connected to a PC. The absolute efficiency of detection is obtained using a standard ^{137}Cs source. The fuel element is axially and radially scanned to obtaine both gross gamma activity profile and gamma activity profiles of some isotopes such as ^{137}Cs, [134]Cs, [95]Zr-Nb, [103]Ru-Rh, [106]Ru-Rh, [140]Ba-La, [144]Ce-Pr. The axial gamma profile enables to check for flux peaking and loading homogeneity.

Figure 6 (a) shows the axial activity distribution of the ^{137}Cs isotope in the fuel element shown in the Fig.2, which has been irradiated in the TRIGA reactor in a power ramping test. A prominent depression of count rate at fuel pellet interfaces is observed, a fact which mean that there is no pellet interaction. The gamma activity profiles of the ^{137}Cs isotope highlight practically a symmetric loading of the fuel element.

A method of tomographic reconstruction based on a maximum entropy algorithm has been developed. This method provides informations on the radial distribution of FPs activity in a cross section of the fuel element. The tomography can be used as complementary method for detection of the sheath defect.

Figure 6 (b) shows, qualitativelly, the tomographic image of the radial ^{137}Cs gamma activity distribution in a central area cross section of the fuel element shown in Fig.2. This tomography indicates that the ^{137}Cs isotope migrated from middle to periphery of the fuel element and redistributed according to the temperature profile.

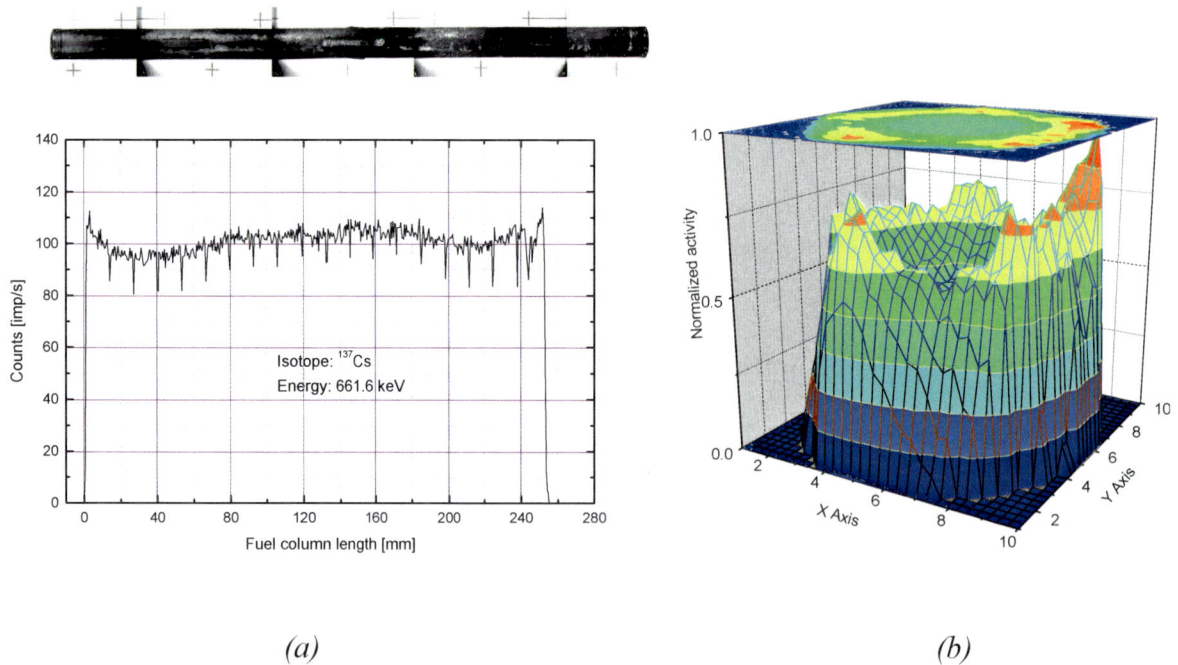

(a)   *(b)*

Fig.6 Axial gamma profile (a) and tomography (b) of ^{137}cs isotope in a CANDU fuel element tested in a power ramping in the triga reactor

2.1.4 Eddy current testing

The eddy current technique provides informations on the integrity of irradiated fuel rod sheath. The testing equipment is composed of a monochannel eddy current flaw detector operated in the test frequency ranging from 1 kHz to 1 MHz and a probe coil. The empty Zircaloy-4 tubes (diameter 13.08 mm, wall thickness 0.38 mm and length 500 mm) having artificially produced defects are used for calibration. The standard artificial defects are: external and internal longitudinal and circumferential notches and holes. Defects of a few hundredths of a millimeter can be detected.

In Fig. 7 is shown the failure signal provided by eddy currents induced in the sheath of a CANDU fuel element irradiated in the TRIGA reactor in an overpower test.

The sheath defects detected by this technique have been confirmed by optical microscopy as shown in the figure.

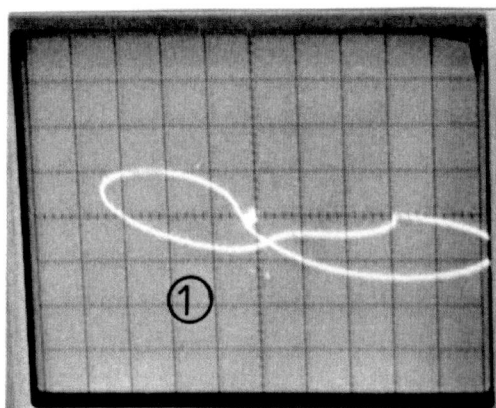

Fig.7 The failure signal provided by eddy current detected by optical microscopy

Fig.8 The sheath defects

2.2 Destructive post-irradiation examination techniques

The nuclear fuel destructive PIE applied in the LEPI hot cells includes the following techniques:

- Fuel rod puncture test;
- Optical microscopy;
- Chemical analysis and burn-up determination;
- Mechanical testing.

2.2.1 Fuel Rod Puncture Test

This technique is used to measure the pressure and volume of fission gas inside the fuel rod and the fuel rod internal void volume. A quadrupole mass spectrometer installed at the outside of the hot cell enables to analyse the fission gas composition.

2.2.2 Metallographic Examination

The microstructural changes of the fuel (fuel restructuring, grain size), the oxide layer thickness, hydriding and hardness of the sheath, fuel-sheath interaction are important parameters in determining how well fuel elements have performed in reactor (see Fig.9 a-f). The examination equipment consist of an optical microscope having a magnification x500 and being equipped with an image analyzer and a Vickers microhardness tester. The chemical etching of the samples reveals fuel restructuring (columnar and equiaxial grains) and sheath hydriding. A computerized analysis system is used for the quantitative determination of structural features, such as grain and pore size distribution.

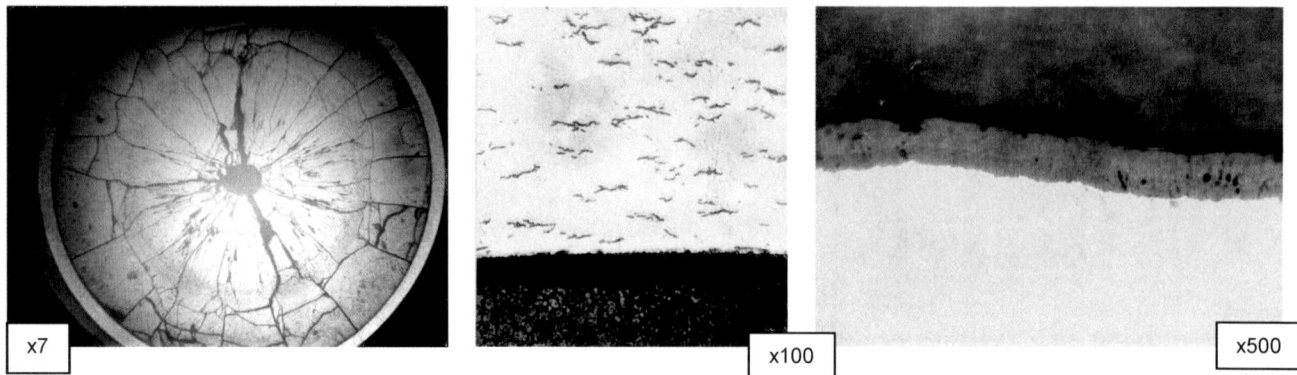

| a) Cross-section | b)   Sheath hydriding | c) sheath oxidation |

d) Columnar grains at section
center

e) Grown equiaxial grains at
radiusmiddle

f) as a sintered grains at
section periphery

Fig.9 Metallographic examination results of a candu fuel element irradiated in the triga reactor in an overpower test

2.2.3 Mass spectrometry

This technique is used to determine the nuclear fuel burn-up that is the energy developed by the unit of mass of the fuel. The procedure includes: fuel rod sample chemical dissolution, isotopic separation, deposition on the filament and determination of isotopic ratios for U: U-234/U-238, U-235/U-238 and U-236/U-238. The method provides an accuracy of ±4 % and is used also to calibrate the rapid gamma scanning method for burn-up determination.

2.2.4 Tensile testing

The mechanical properties of the irradiated fuel element sheath such as fracture strength, creep etc. can be determined by this technique. The testing equipment installed into the hot cell consist of a tensile-testing machine of 50 kN equipped with a furnace. Available test conditions include a temperature range from $20^{o}C$ to $1000^{o}C$ under air atmosphere. Crack growth monitoring and data acquisition and handling are fully computerized.

3.   CONCLUSIONS

The post-irradiation examination of CANDU fuel elements manufactured at Pitesti and tested in the TRIGA material testing reactor has been performed in the hot cells from ICN Pitesti since 1984 as part of the Romanian research programme for the manufacturing, development and safety of the CANDU fuel.

The results obtained by non-destructive and destructive examinations concerning the integrity, dimensional changes, oxidation, hydriding and mechanical properties of the sheath, the fission products activity distribution in the fuel column, the pressure, volume and composition of the fission gas, the burn-up, the isotopic composition and structural changes of the fuel have enabled the CANDU fuel behaviour characterization after its testing in the TRIGA reactor both in normal operation and in accident conditions.

During 20 years of post-irradiation examination, the Post-Irradiation Examination Laboratory from ICN Pitesti has obtained the operational and professional ability to evaluate the performance of the CANDU nuclear fuel from Cernavoda NPP.

Irradiation devices for fission and fusion materials testing in the HFR, Petten

B. VAN DER SCHAAF, J. VAN DER LAAN

NRG, Petten, Netherlands

Abstract: The High Flux Reactor, HFR, in Petten is a 45 MW pool type materials test reactor. The HFR reaches about 300 full power days per year. Peak radiation damage levels of 6 displacements per atom can be generated in steel per year in-pile. The core has experiment positions have a diameter about 70 mm. The effective core height is about 500 mm in length. The support for fission technology of existing and future power plants consists of capsules containing samples to be conditioned in environments relevant for: light water reactors, LWR's, Liquid Metal Reactors, LMR's, and Very High Temperature Reactors, VHTR's. Several examples of fission relevant materials irradiations will be presented and discussed. Static capsules provide means to irradiate test samples in helium, carbon dioxide, CO_2, water, sodium or liquid lead in the temperature range of 200 $^\circ$ C up to 1300 $^\circ$ C. The post irradiation testing of the materials can be performed in the NRG hot-cell laboratory on the Petten site. There are wide ranges of mechanical, physical and chemical test apparatuses available to satisfy the research and development needs. In-pile materials' testing is provided in the form of in-pile relaxation and creep devices and in-pile corrosion in water, lead and CO_2 N, some will be highlighted in more neutron flux transients are used to simulate reactor transients. The materials tested include both structural materials and fuel and reflector materials. The fuel focus is on UO_2, but considerable effort is put into inert matrix material development for the burning of actinides. Steel research is concentrating on reactor pressure vessel steels for new generations of VHTR's. Very special steel applications are studied in rigs spectrum tailored in a way that they resemble the environment for windows of lead cooled particle beam controlled fission rectors. Fusion power, such as generated in ITER, requires special materials and components to extract the tritium and fusion energy in a safe and reliable way. In the HFR, Petten, both materials and components are subjected to environments relevant for application in ITER and subsequent fusion power plants. The materials and components needed near the plasma such as the first wall and breeding blankets are developed with rigs in the HFR. Tritium breeding modules are tested in static and transient conditions in-pile, which produces data relevant for ITER and power plants in the future.

1. INTRODUCTION

The High Flux Reactor, HFR in Petten is a 45 MWth pool type materials test reactor. The major tasks are the production of isotopes for medical applications, and research and development of materials and components. Peak radiation levels of 8 dpa in steel can be reached per year. These levels are adequate for isotope production. This radiation damage rates is sufficient for the majority of the research and development of new materials and components for the new generation fission reactors, GEN-4 and fusion energy devices. For the end-of-life conditions of fast reactors and fusion power plants additional materials testing devices are necessary such as IFMIF for the irradiation with 14 MeV neutrons highly relevant for the components near the plasma with deuterium-tritium reactions.

The experimental programmes are not restricted to materials testing alone. Increasingly the HFR irradiates relevant sections of components to be applied in fission reactors and fusion devices. The results of the component testing are important for the quantification of the interaction of structural, functional materials and coolants. Also the process parameters a be verified with the in-pile operation of sub-components. This paper highlights the major projects presently in progress in the HFR. The highlights concern materials and component development for future reactors, but several examples of results relevant for light water reactors now in operation are also included.

The HFR, Petten, started operation in 1961 and had a major overhaul in 1984. The development of a new generation of fission reactors for this century and the fusion power plant development necessitate to rejuvenate the HFR. The PALLAS initiative investigates the design and construction of the successor of the HFR based on the foreseeable power reactor requirements for the first half of the 21st century. The need for at least doubling the neutron flux of the PALLAS operation is addressed in this paper on the basis of a conceptual design. The need for a flexible core configuration in PALLAS for the economy of the operation is addressed in the paper.

2.   HIGH FLUX REACTOR, HFR, PETTEN

The High Flux Reactor, HFR, in Petten is a 45 MW pool type materials test reactor. It started operation in 1961. The reactor vessel after reaching about 100dpa peak was replaced by a newly designed one in 1984 with a design life into the 2020's. The HFR reached about 300 full power days per year over the last decennia. Peak radiation damage levels of 6 displacements per atom can be generated in steel per year in-pile. The core has experiment positions with a useable diameter of about 70 mm. The core scheme is shown in Fig. 1.

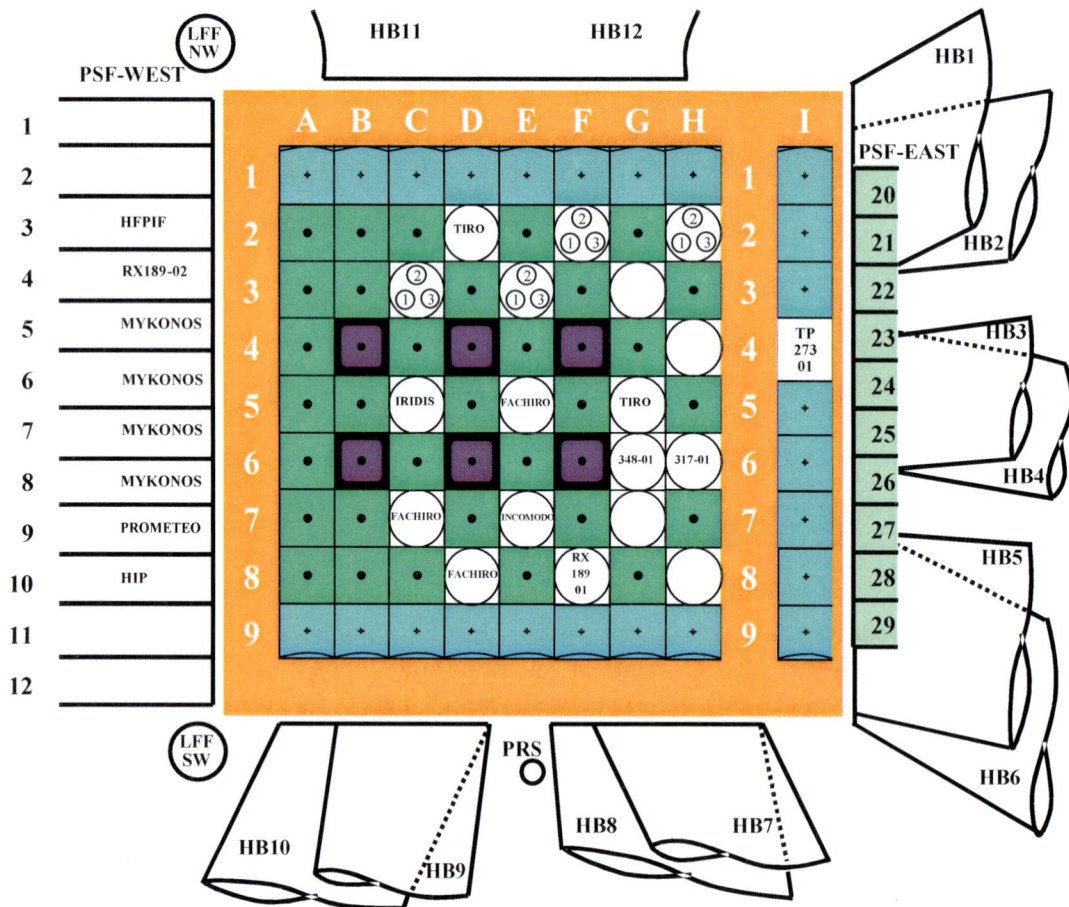

Fig. 1: Scheme of the HFR, Petten core with the Pool Side Facility, PSF, to the left and the beam majority to the right of the HFR core. The circles indicate experiment positions in the core

The effective core height is about 500 mm in length. The reactor power is 45 MW thermal. The peak neutron fluxes amount to:

Thermal neutrons: $1.6*10^{18}$ n.m^{-2}.
Fast neutrons: $2.5*10^{18}$ n.m^{-2}.

At present the HFR has four main research and production applications:

- Technology development and demonstration for new power plants (Gen-2,3 & 4, Fusion)

- Reduction of nuclear fuel waste storage time (factor 1000 reduction)
- Pharmaceutical radio-isotope productions (60% of production to hospitals in the EU)
- Beam tubes for materials science and engineering, health treatment with boron neutron capture.

3. MATERIALS RESEARCH FOR NUCLEAR POWER PLANTS

Table 1 provides the fission and fusion relevant examples of materials irradiations . Static capsules provide means to irradiate test samples in helium, carbon dioxide, CO_2, water, sodium or liquid lead in the temperature range of 200 ^{o}C up to 1300 ^{o}C. The post irradiation testing of the materials can be performed in the NRG hot-cell laboratory on the Petten site. There are wide ranges of mechanical, physical and chemical test apparatuses available to satisfy the research and development needs.

TABLE 1. FISSION AND FUSION MATERIALS RESEARCH IN THE HFR, PETTEN

Material	Fission: Gen-4, HTR	Fusion: ITER, DEMO
Structural	Pressure vessel steel Canning steel nano-microstructure. Graphite, composite	Low activation steels ODS steels Tungsten SiCSiC ceramic composites
Functional	Fuel particle elements Inert Matrix Actinides	Lithium ceramics Beryllium pebbles
Process	Fuel element test in all conditions Pb-Bi compatibility Graphite creep	Tritium release Pb-Li behavior Bolt relaxation First wall simulation

Radiation damage in materials depends strongly on the neutron spectrum that the materials experience under operational conditions. The thermal fraction can bring about nuclear reactions producing isotopes strongly affecting the physical and mechanical properties, for example helium production from boron in steel. The HFR core and PSF offer neutron spectra quite different in contributions from the thermal and fast spectrum tail. Simulating the environment for the application sometimes needs more spectrum adjustment than the irradiation positions offer. In several experiments the thermal part of the neutron spectrum is further tailored using cadmium or hafnium shields [1]. The shields greatly reduce the thermal neutron fraction reaching the samples under irradiation.

3.1 Fission power plant applications

The support for fission technology of existing and future power plants consists of capsules containing samples to be conditioned in environments relevant for: light water reactors, (LWR), Liquid Metal Reactors (LMR), and Very High Temperature Reactors (VHTR). In-pile materials testing is provided in the form of in-pile relaxation and creep devices and in-pile corrosion in water, lead, lead bismuth and CO_2.

The materials tested include both structural materials, and fission fuel, and reflector materials. The fuel focus is on UO_2, but considerable effort is put into inert matrix material development for the burning of actinides. In the EU considerable effort is put into the development of fuel cycles that allow the utilization of plutonium stockpiles for power generation in LWR and LMR's. The application for LWR requires the Pu to be incorporated as Fuel in an Inert Matrix, IMF. The approach uses a complete chain of analyses for prediction of IMF behavior in an LWR core, fabrication of test batches, irradiation in the HFR and Post-Irradiation

Examination (PIE), to verify the modeling basis for the analyses. In the HFR advanced fuel matrices have been irradiated leading to results supporting the analyses [2].

Steel research for new reactors of Generation-4 is concentrating on reactor pressure vessel steels, such as the advanced modified 9Cr 1 Mo steel, for new generations of VHTR's. Neutron irradiations under static temperatures of 370 $^{\circ}$ C nominal with less than 10 degrees deviation are carried out. Later the irradiated samples are mechanically tested under tensile and creep conditions relevant for both normal and abnormal condition. In particular the creep property measurements are highly relevant for accident analyses.

For existing reactors the effects of Irradiation Assisted Stress Corrosion Cracking, IASCC, are studied. In order to determine the effects of irradiation on weld residual stresses welds as large as possible are irradiated. Before and after the neutron irradiation in the HFR to several neutron dose levels the PIE stress corrosion properties and weld residual stresses are measured [3, 4]. The results are used for the prediction of the IASCC in welds in core shrouds manufactured in the past from Type 304 and Type 316 steel. One of the results is that the fracture toughness of the Type 304 and 316 welds decreases by a factor 3 after 0.3 to 1.0 dpa, but the toughness still remains above 100 $kJ.m^{-2}$. In the reference condition the scatter of the results amounts from 25 to 35 %, brought about by the crack path through a limited number of welding passes. After neutron irradiation the scatter is reduced to less than 10 %: the radiation damage has become the dominant influence on weld toughness.

In the third quarter of the last century much effort was devoted to the development of graphite for moderating and fuel element purposes in gas-cooled reactors. Analyses carried out in the GEN-4 programme show that the gas-cooled reactors have several inherent properties which make them important candidates for the commercial generation of both electricity and process heat in this century. The graphite developments completed in the last century are of limited value for the present design activities. The sources for the graphite production have changed and the processing for nuclear grade graphite as well. Therefore modern graphite development is supported with irradiation programmes in the HFR, Petten. Major issues are the dimensional stability of graphite under irradiation, heat transfer, and strength. Graphite shrinks and grows dependent on the radiation damage level, and the conductivity decreases due to radiation damage. The graphite is irradiated in the HFR at temperatures of 750° C and 950° C, highly relevant for the operating conditions for HTR and VHTR . The post-irradiation measurements include physics (density, conductivity) and mechanical properties (elastic modulus, strength) measurements [5]. Differences with the conventional graphite have been observed and will have to be incorporated in the conceptual core designs of the GEN-4 HTR's.

Very special steel applications are studied in rigs spectrum tailored in a way that they resemble the environment for windows of lead-bismuth cooled particle beam controlled fission rectors. Such Accelerator Driven Systems, ADS, might play a role in burning actinides in the future. Actinide burning is a method for reducing the storage time of burnt fission fuels. The lead bismuth corrosion effects on the chromium steel (intended for application in pressure vessel and fuel canning) are the main objective of the HFR irradiations. The formation of polonium by neutron reactions, mainly from the bismuth nuclei, and its' influence on the corrosion behavior is another goal of the investigation.

4.   FUSION MATERIALS RESEARCH AND DEVELOPMENT

Fusion power development is presently mainly aiming on the sustained reaction of deuterium with tritium in hot plasma. In addition the inertial confinement approach, where deuterium tritium containing pellets are ignited with an intense external source, is pursued on a more limited scale. Both approaches depend on tritium production of neutron interaction with lithium in breeding blankets. The blankets, be they next to burning plasma or burning pellet stream, require special materials and components to extract the tritium and fusion energy in a safe and reliable way. In the HFR, Petten, both materials and components are subjected to environments relevant for application in fusion power plants. The plasma experiment JET in Culham has produced MW's of fusion power though for a short time. The next large international, plasma experiment, ITER in Cadarache to be operational in 2015, will produce 500 MW of fusion energy for periods of at least 500 s. The materials and components needed near the plasma, such as the first wall and breeding blankets in ITER, are developed amongst others with rigs in the HFR, Petten. Tritium breeding modules are tested in static and transient conditions in-pile, which produces data relevant for ITER and power plants in the future. These data include the behavior of the functional materials lithium ceramics and lithium lead and the structural materials steel and advanced silicon carbide ceramic composites.

One of the irradiation projects in progress concerns the relaxation of nickel base alloy bolts used for fastening components in the ITER vacuum vessel. In a neutron field alloy 718 shows a tendency for relaxation. The magnitude of the phenomenon and the mechanisms controlling it are important information for the designers of the bolted structures. In the HFR pre-stressed bolts and springs made of different materials with ranges of heat treatments are subjected to increasing levels of radiation damage. After irradiation the relaxation effect is determined and the results are compared with the tendencies predicted [6], Fig. 2.

It can be observed that after about 0.7 dpa the pre-stress level in the bolt material can be halved. The experimental work and modeling show that material composition and heat treatment selection can optimize the required properties such that the ITER requirement can be fulfilled.

In fusion power plants replacement of breeding blankets interferes with the ideal continuous operation of the plant. Outages due to blanket replacements must thus be minimized. This required highly reliable materials and manufacturing methods for the breeding blankets. Austenitic steels suffer from helium embrittlement and swelling, whereas ferritic steels are not. In addition the masses of materials, needed for the blankets, should produce low radiation levels to reduce waste costs. In the ideal case re-use of the structural materials should be possible. Reduced activation ferritic martensitic, RAFM, steel should just do that. The main constituents iron, chromium and some tungsten and the very low level of impurities obtained assure the recycling potential. At present they are fabricated on industrial scale, but their composition, heat treatments and manufacturing practices deviate considerably from conventional steels. An experimental programme to determine the properties and understand the failure mechanism is needed to provide designers with facts that can convince licensing authorities of their use in breeding blankets. In ITER, their operation will be demonstrated, which requires experimental and modeling evidence. The irradiation programme in the HFR, Petten, contributes in the form of neutron radiation of RAFM steels to dose levels relevant for the end of life condition of ITER breeding blanket test modules. NRG has focused on the irradiation effects of the different types of welding techniques, [7], anticipated for fabrication of blankets with RAFM. Figure 3 gives a selection of results for the fracture toughness of the different welding techniques.

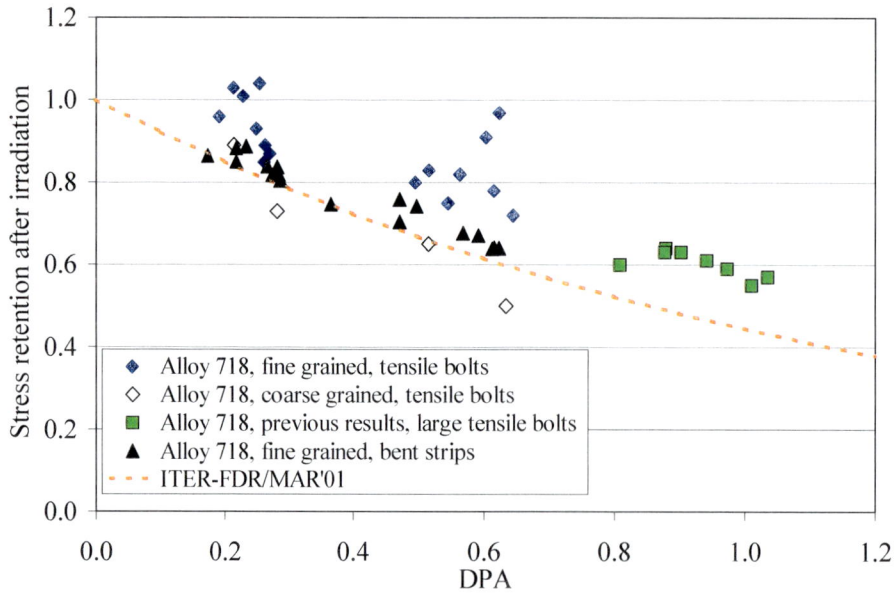

Fig. 2: Stress relaxation of alloy 718 after irradiation in the HFR, Petten, to the damage level indicated in displacements per annum, dpa.

In the future the efficiency of fusion power plants can increase using more advanced high temperature breeding blankets using silicon carbide ceramic composites. Also this material is of a low activation nature and can operate at temperatures well over $1000\ ^\circ$ C. The radiation resistance at present is limited and research and development aim for increasing heat conductivity and mechanical properties. In an irradiation programme including rigs with operating temperatures up to $1000\ ^\circ$ C the materials developed in industry are subjected to neutron radiation and subsequently tested for the irradiation effects on physical and mechanical properties. Major results [8] indicate that the silicon carbide matrix must be improved to raise the prost-irradiation properties to values acceptable for the designer.

At present reductions in heat conductivity by a factor of 10 still occur which promotes their application as heat barrier, but not as a structural material. This limitation must be overcome to allow for higher thermal efficiencies of fusion power plants.

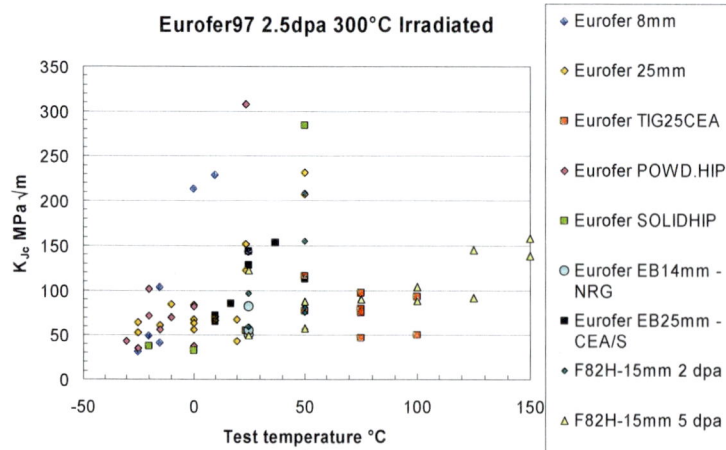

Fig. 3: Fracture toughness of different weldings as dependent on the testing temperature

5.   THE PALLAS INITIATIVE

The PALLAS specification is aiming for both innovation, and continuity of a safe and reliable neutron irradiation capacity on Petten site. The initiative is intended as an integral part of the EU nuclear research infrastructure [9]. The new isotope and materials test reactor definition of the PALLAS specification is a four party initiative:

1. Joint Research Centre – Institute for Energy, Petten
2. Mallinckrodt Medical, Petten
3. Technical University Delft.
4. Lead by NRG, Petten

The year 2015 is the target start date for the PALLAS successor to the HFR, Petten. The end-of-life of the HFR Petten is not fixed in that year, but it is expected that the end of the operation of the HFR is around 2020. Next to the four party initiative other stakeholders involved in decision making, such as local authorities and population are involved.

The PALLAS specification has as key elements of choice flux and volume:

- Peak fast flux: 5×10^{18} ($\sim 2 \times$ HFR),
- Peak thermal flux: 5×10^{18} (~ 2-$3 \times$ HFR).

The reactor power is expected to be in the range of 40-80 MW, controlled by the doubling of the neutron flux and the flexible core configuration. Fuel cost front and back end for the reactor operation is an important consideration controlling also the thermo-hydraulic design and plant life target. Flux shaping is considered an important tool to economize on the PALLAS operation. Squeezing the core to reduce the neutron leakage is needed to limit the irradiation volume and thus the unnecessary burning of fuel. Combining isotope production with materials and components testing seems essential for efficient and economic operation of PALLAS.

Flux boosting in particular experiments is contemplated, to satisfy specifications for high fluxes locally. Optimization is foreseen to be determined by potential supplier(s) of the reactor detailed design. Technology scans must provide key innovation parameters. Proven technology will be essential in assuring the acquisition of the licenses for the PALLAS isotope, materials and component test reactor.

6.   CONCLUSION

1. The four main HFR utilizations: technology development for Gen-4 and fusion, reduction of nuclear fuel waste storage time, pharmaceutical radio-isotope productions, and beam tubes for materials science and engineering show the importance of research combined with isotope production.

2. The examples shown illustrate that the HFR provides environments relevant for the next generation fission reactors and fusion power plant development.

3. Besides materials conditioning and testing the in-pile testing of components and sub-assemblies, providing process data and materials interaction effects, grows significantly.

4. The preparation for the HFR successor, PALLAS, focuses on doubling the neutron flux, and flexible core configuration. In this way the requirements of PALLAS users for the first half of this century can be satisfied.

REFERENCES

[1]   RENSMAN, J., et.al., *"NRG SPIRE Contribution: mechanical test results from MTR irradiation up to 3 dpa @ 250°C"* 20564/04.62317/P, Petten, 11 October 2004

[2]   den EXTER, M.J., NEUMANN, S., TOMASBERGER, T., Immobilization and behavior of Tecnetium in a Magnesium Titanate Matrix for Final Disposal, MRS2005 Symposium " Scientific Basis for Nuclear Waste Management", (accepted for publication in 2006).

[3]   van der SCHAAF, B., et al. Effects of neutron radiation and residual stresses on the corrosion of welds in light water reactor internals, ENC 2005 Proceedings paper no. 19.

[4]   OHMS, C., page 69 of this publication

[5]   van der LAAN, J. G., ARJAN VREELING, J.A., Graphite irradiation testing at NRG Petten, Paper presented at the Conference on Ageing Management of Graphite Reactor Cores (AMGRC 2005), held at Cardiff University Hall Conference Centre, United Kingdom, November 28-30, 2005, submitted for publication in the proceedings.

[6]   SCHMALZ, F., Irradiation induced stress relaxation results for Alloy 625+ and PH13-8Mo, NRG report 20726/05.64920/P, May 2005.

[7]   RENSMAN, J.W., NRG Irradiation Testing: Report on 300 ° C and 60 ° C Irradiated RAFM steels, NRG report 68497/P, 30 August 2005.

[8]   HEGEMAN, J.B.J., JONG, M., PIERICK, P.T., van der LAAN, J.G., High temperature tests of 2D and 3D SiCf/SiC composites, ICFRM-12, Santa Barbara, December 2005, submitted for publication in the proceedings.

[9]   van der SCHAAF , B., BERGMANS, D., van der LAAN, J.G., PALLAS Reactor Project, ESFRI expert group meeting, Presentation in Brussel, December 2, 2005

Research reactors in Argentina and their neutrons applications

M. SCHLAMP

Projects Coordination Unit of Nuclear Reactors and Power Plants CNEA – Argentina

Abstract: Argentina has been successfully developing nuclear energy applications for many years. In particular neutron applications have been developed and used in its research reactors and accelerators. The Argentinean Atomic Energy Commission (CNEA) has six Research Reactors, three of them are Critical Assemblies.

The main objectives of these Research Reactors in CNEA's strategic plan are:
- Personnel education and training for building, start-up, operation and maintenance of Atucha II
- Support for NPP life extension (Embalse, Atucha I)
- Neutronic parameters measurements as support of CAREM design project
- 90% production of national radioisotopes utilization for medical purposes
- Neutrongraphy and prompt gamma capture techniques analysis
- Different techniques of BNCT applied to humans and animals
- under and post graduate students education
- training for national and international operators
- Public diffusion about the benefits of nuclear applications
- neutron activation analysis for aero spatial and nuclear industry
- irradiation of national developed and constructed silicide fuel for qualification to export
- RR and NPP instrumentation qualification

1. INTRODUCTION

1.1 CNEA structural relationships

As shown in fig. 1, two of the six Research Reactors from CNEA are operated by national universities (NUC and NUR). The Nuclear Power Plants are operated by Nucleoelectrica Argentina S.A. (NA-SA) which depends, as well as CNEA, from the Energy Secretariat. The regulatory body (ARN) depends from the General Secretariat.

Figure 2. shows the geographical distribution of the research reactors in Argentina. RA-0 is located in the National University of Rosario, RA-1 is in Constituyentes Atomic Center (Buenos Aires), RA-3 is in Eziza Atomic Center (Buenos Aires), RA-4 is in the National University of Rosario, RA-6 in Centro Atomico Bariloche and RA-8 in Pilcaniyeu Technological Center (Bariloche).

2. RESEARCH REACTORS CHARACTERISTICS

2.1 RA-0: operational

Critical Assembly
Criticality   : 1970
Place   : Cordoba National University
Power   : 1 W
Type   : Tank
Utilization   : Teaching, training and research
Fuel   : UO2
Fuel Element   : Rods
Enrichments   : 20 %
React. XS   : 0.4 $

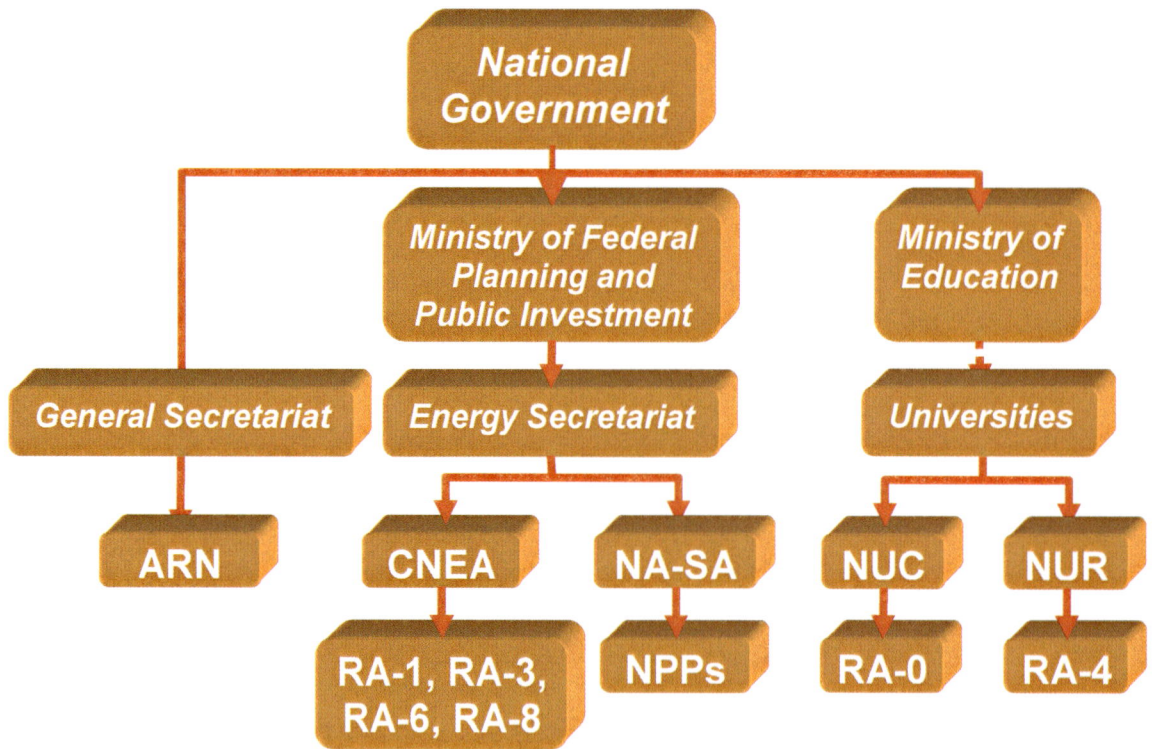

Fig. 1. Relationship between CNEA and other nuclear institutions

Fig. 2. Geographical distribution of the research reactors in Argentina

2.2 RA-1

Critical Assembly: In operation

Criticality   : 1958.
Place   : Constituyentes Atomic Center
Power   : 40 kW
Type   : Tank
Utilization   : Research, training, BNCT, material testing
Fuel   : UO2
Fuel Elements : Rods
Enrichments   : 20 %
React. XS   : 1.5 $

2.2.1 Irradiation facilities for material testing

- Two vertical channels in the core center, with thermal Flux of 4×10^{11} n/(cm^2 s) and fast flux of 1×10^{12} n/(cm^2 s)

- One High Temperature Irradiation Facility (100 $^\circ$ C to 400 $^\circ$ C) for materials studies under irradiation at a selected temperature and inert atmosphere (He, Ar, N)

- One facility at low temperature for materials studies under irradiation at nitrogen or helium liquefaction temperatures, in order to avoid the migration of defects produced during the irradiation

2.3 RA-3: *Operational*

Research &Production Reactor

Criticality   : 1967.
Place   : Eziza Atomic Cent.
Power   : 10 MW
Type   : Tank
Utilization   : Radioisotopes production, research, material testing
Fuel   : UO2
Fuel Elements : MTR
Enrichment   : 20 %
React. XS   : 8 $

2.3.1 Irradiation facilities for material testing

- 4 sample irradiation boxes inside the core (each has 16 irradiation places, Φth: 4×10^{13} to 1.2×10^{14} n./(cm^2.s)

- special irradiation box for mini-plates targets, placed in the center of the core (12 mini-plates could be placed inside, Φ th $= 2.4 \times 10^{14}$ n./(cm^2.s))

Irradiation facilities for material testing:

- High Temperature Irradiation Facility (for materials studies under irradiation at temperatures from 100 $^\circ$ C to 500 $^\circ$ C)

- Irradiation Facility at reactor cooling temperature

- Six Radial Irradiation conduets : In one of them a neutron radiography facility worked until 1988, and it is going to be renewed to be used next year.

2.4   RA-4: *operational*

Critical Assembly

Criticality   : 1971.
Place   : Rosario NU
Power   : 1 W
Type   : Homogeneous
Utilization   : Teaching, training and research
Fuel   : UO2
Fuel Elements : Polyethylene plates
Enrichment   : 20 %
React. XS   : 0.4 $

2.5   RA-6: *Operational*

Criticality   : 1982: operational
Place   : Bariloche Atomic Center
Power   : 500 kW
Type   : Tank
Utilization   : Research, training, NAA, BNCT
Fuel   : UO2
Fuel Elements : MTR
Enrichment   : 90 %
React. XS   : 2 $

2.5.1   Irradiation facilities

- 5 Horizontal tubes (2 in use); maximum thermal flux: 1×10^{13} n/(cm^{2}sec)

- BNCT beam

- 1 Pneumatic transport/transfer system

2.5.2   Under development

- HEU core conversion to LEU (start-up beginning 2007)

- Power upgrade, from 500 kW tp 3 MW

- On line Neutron Radiography (2007)

- Prompt Gamma Capture analysis device (2007)

2.6   RA-8: Under extended shut down

Critical Assembly
Criticality   : 1998.
Place   : Pilcaniyeu Technological Center
Power   : 10 W
Type   : Tank
Utilization   : CAREM fuel test
Fuel   : UO2
Fuel Elements : Rods
Enrichment   : 1.8 & 3.4 %

3 PRESENT EXPERIMENTS FOR MATERIALS TESTING

3.1 Irradiation of diffusion couples U-Mo alloys/ Al-Si alloys

Out of pile experiments have been done at 550°C and 340°C using Al/Si alloys containing 0,6 to 7 % Si. Reported results were considered promising concerning the effect of Si in changing the composition of the interaction layer with respect to the case of pure Al; this may contribute to solve the failures in irradiated fuel elements. The changes were that (UMo) (Al-Si)3 is now the major component of the interlayer and (UMo)Al4 is not detected.. This last phase was associated to the porosity found in the qualifying post irradiation experiments.

The objective of *in-pile experiments* is to characterize the evolution of the interdiffusion zone with the neutron fluence. Diffusion couples for the irradiation experiments consist in two plates of the Al alloy encasing an U-Mo foil. They were joining together by Friction Stir Welding (FSW) technique, which has proved to be very suitable technique as in the fabrication of monolithic miniplates. The alloys used were U-7w%Mo or U-9w%Mo made from depleted U, and 0.7wt% Si -Al alloy. First irradiations are in course in the RA-3 reactor. The selected position is in the centre of the core, in a box where the thermal flux is $2x10^{14} n/cm^2 s$. It is planned to irradiate from one to three month in periods of five days. Post irradiation analysis in hot cells are planned to start in June 2006.

3.2 Irradiation embrittlement in NPP pressure vessel's steels

The objective of the experiment is the determination of the lead factors effects in the mechanic behavior of NPP pressure vessel's steels. Recommended factors for surveillance programmes of water cooled pressure vessels are in the range of 1 to 3. Tests carried out until now have factors up to 800, which does not guaranty that the irradiation damage is similar to the one suffered by the pressure vessels in service. Fracto-mechanical probes are being irradiated in RA-1 research reactor with different lead factors to determine embrittlement parameters. Working temperatures will be around 300 °C. Two irradiations will be carried on: 36 days with maximum flux; and 360 days with one tenth of maximum flux, to reach the same dose.

3.3 Synergic effect between hydrogen content and irradiation damage in Zr alloys

The main objective of this work is the determination of the endurance low due to irradiation of Zr alloys with hydrogen. In a power reactor core the incorporation of hydrogen and the irradiation damage occur simultaneously. Zr and Zry probes have been irradiated in RA-1 research reactor with different hydrogen contents obtaining an important change in the damage recovery temperatures as a function of hydrogen content.

4 FUTURE NEEDS

4.1 High pressure and temperature fuel irradiation device

Argentina produces fuel rods for Atucha I, Embalse and CAREM (prototype) nuclear power plant. There is a project under development to construct a high pressure and temperature fuel irradiation device to simulate nuclear power plant fuel behavior in normal conditions, power ramps and with neutron flux higher than $10^{14} n/(cm^2 s)$.

The characteristics of the loop are:

- Light water loop in NPP operating conditions (~12 Mpa and 310°C)

- capable for 3 fuel rods (CANDU, ATUCHA, CAREM, CARA, etc)

- length: 40 cm

- instrumented rods

- maximum linear power: 600 W/m

- maximum power: 70 kW

- placement in reactor: reflector zone

There is an evaluation to place the irradiation device at RA-3 research reactor with positive results.

4.2 Replacement multi purpose research and production reactor

Argentina's main radioisotop production reactor, RA-3, is almost 30 years old and will need to be replaced in the next decade. The first question that arises is: which kind of reactor should be the new one?

Regarding nuclear fuel for research reactors, CNEA produces fuel plates for RA-3, RA-6, OPAL (Australia) reactors and fuel rods for RA-1 reactor. It also produces targets for fission Mo99. Such fuel elements should be tested in a Research Reactor with a suitable neutron flux.

There is also a need of analysis devices for: fresh fuel Neutron Radiography and irradiated fuel Neutron Radiography. Also a Neutron Diffractometer is observed as a very useful tool for research and for the industry.

5.   EVALUATION OF A NEW RESEARCH REACTOR

The need of the new reactor and its characteristics are been discussed since several years in different forums and meetings like:

2003: Argentina participated in a regional "Technical Meeting on Strategic Planning (SP) for Research Reactors Utilization" organized by Mr. Shirinwas Paranjpe.

2004: a Strategic Plan was prepared for each RR in Argentina.

2005: a "National Workshop on SP for RR" was organized in Argentina and its output was a "National SP for the RR and Production Subprogramme for Argentina".

    - one of the outcomes was the decision to design and construct a new Research and Production Reactor to replace RA-3.

6.   ALTERNATIVES ANALYZED:

- RA-3 upgrade

- New Radio Isotopes Production Reactor

- New Multi Purpose Reactor

After preliminary evaluations and analysis, taking into account by one hand the needs and the possible budget of the country and by the other hand the cost and potential uses of each alternative, a new multi purpose reactor seams to be the best option.

The next steps include a deeper stackeholders assessment in the country and outside the country and the preparation of a strategic plan for the future reactor to evaluate the strength and weakness of the project.

Some considerations that will be taken into account for the strategic plan are: the support offered by the Energy Secretariat and by the Science & Technology Secretariat, the interest from the private industry and from the Region to participate in the project and the services of the reactor.

7. CONCLUSIONS

There is a need in Argentina to count with several neutron beam facilities to cover important aspects related with industrial applications, scientific and technological development and the design and construction of Research and Power Reactors.

A new Research and Production Reactor will be constructed to replace RA-3 taking advantage of the experience in Argentina to design and build research reactors

It will be a Multipurpose Research Reactor, with specific characteristics that will depend on a balance between the stakeholders and the budget

Regional participation is welcome

Materials research — A challenge for fission, fusion and accelerator research

G. MANK

International Atomic Energy Agency, Vienna

Abstract: Different topics of materials research are of interest to the energy sector, fusion research and basic research at accelerators. The Physics Section supports and promotes these endeavors on demand and recommendation by the Member States. The paper will briefly describe topics covered by the Physics Section related to materials research at Research Reactors and for energy related applications, research at accelerators and research for the next generation of fusion devices

1.   INTRODUCTION

Since more than 60 years research reactors have been used for materials research in basic science and in energy related applications[1]. The global population of research reactors had peaked in the 1970s at 400 but is about 270 now. With this reduced number of facilities there is still much research using neutrons that needs doing. In the 1980s' the possibilities of spallation sources, with much higher neutron peak fluxes than research reactors, have been investigated. With large spallation sources coming soon on-line such as SNS, new research can be pursued.[2] The target of a spallation source can be a challenge if e.g., a mercury target is introduced. High neutron fluxes delivered by spallation sources will be used for the investigation of wall and blanket materials for future fusion reactors. The demands for the blanket and structural material for fusion reactors are high, as the amount of radioactive waste resulting from the operation of a fusion power plant should be minimal.

The paper will describe some basic points of materials investigation using research reactors, accelerators and for fusion respectively. The topics presented will reflect only a part of the work of the IAEA as supported by the Physics Section.

2.   MATERIALS RESEARCH AT RESEARCH REACTORS

One of the major applications of research reactors is in the production of radio-isotopes, which are used in hospitals in diagnostics and treatment of cancer. Further to this the main objective of the IAEA's programme on materials research at reactors is related to basic science on materials development for the nuclear fuel cycle and research reactor application. With respect to the programme planning for 2008-09 the rational is to do research on advanced materials for NPP, where the requirements are high temperature, high flux, and fast spectrum. The investigations at research reactors include materials for accelerator driven systems (ADS) for transmutation studies, and materials for fusion reactors and science. Performance indicators as indicated are mainly improved conversion from HEU to LEU and fuel tests. Research at reactors should lead to new and improved structural materials.

Further to this, materials will be investigated using standard methods as e.g., neutron radiography. In many cases nuclear applications develop their full potential if they are applied in a complementary manner. For example, combining X ray and neutron radiography can lead to new understanding of structures. The advantage of neutrons is that they are sensitive to many light elements, whereas X rays are more sensitive to heavier elements e.g. the components of steel. Using neutrons one can visualize glue within the metal sheet of a car or plane. Even for arts and archaeology (cultural heritage), neutrons are important as the composition and changes of a painting can sometimes be analysed only by neutrons as they give a different picture for different kind of paints.

Neutron activation analysis is an important technique for elemental analysis in water, air, soil, fish, meteorites, rocks, and even agricultural products and plants. The samples are irradiated in a reactor and later the spectrum of emitted gamma radiation is analysed.

3. MATERIALS RESEARCH AT / FOR ACCELERATORS

Research in materials science using accelerators offers a broad spectrum of activities that builds a cadre of trained experts in Member States, and generates knowledge for innovative methodologies and tools.

The coordinated research efforts underway on ion beam techniques and pulsed neutron sources will lead to new initiatives in materials research of relevance for both the nuclear and non-nuclear fields. Material science studies with the use of accelerators, neutron beams and other nuclear analytical methods are relevant to advanced reactors, nuclear fuel cycle needs and fusion research. The close coordination with NENP in the area of Accelerator Driven Systems (ADS) and with NAAL in nuclear instrumentation for applications in agriculture, health and environment, will bring synergistic strength into programme implementation. The accelerated ageing studies on materials of interest for the nuclear energy programme will be another area of collaborative pursuit together with the Department of Nuclear Energy.

Neutron applications and life science and materials research will be rapidly growing in the future. In addition to basic research new neutron sources and facilities, suitable for many member states, will be needed for education, training, and preparation for industrial applications. Technical Meetings on development opportunities for small and medium scale accelerator driven neutron sources are foreseen to exchange expertise and support common activities in the member States related to the use of pulse neutrons. New ideas evolve from the Technical Meetings, as e.g. the request to increase cold and ultra-cold neutron flux by inserting cold hydrogen or methane moderators. Again the method, which will improve the future research, is under investigation itself at the moment. The new applications related to neutrons are presented to young researchers at schools on pulsed neutrons sources and on accelerators, as organized in close collaboration with the International Centre for Theoretical Physics (ICTP). A new Coordinated Research Project (CRP) on "Improved production and utilization of short pulsed, cold neutrons at low-medium energy spallation neutron sources" is open to Member States within the period 2006 – 2010. The main objective is to enhance the utilization possibilities of low and medium energy spallation neutron sources for research and development in neutron science and applications, by increasing neutron supply at sources and by improving optimum use of neutron techniques in interested Member States. This CRP will be focused on specific research objectives:

- To improve neutron beam flux via development of cryogenic moderators.
- To improve neutron beam resolution via study of new collimator, focusing devices, and mono-chromators.
- To develop compact small angle neutron scattering (SANS) instruments, with micro-focusing properties, for simultaneous installation at neutron beamline.
- To improve capability of strain determination through time-of-flight neutron reactions: experiment and user friendly implementation of the methodology to extract the desired information from the measured cross-sections.
- To carefully consider how proposed techniques and improvement could be made available for interested Member States.

The trend in neutron research is shown best in Fig. 1, where it is shown, that the neutron peak flux using even the best available Research Reactors as ILL Grenoble is in the order of 10^{15} n/cm^2/s, whereas the new build American Spallation Neutron Source (SNS) will have peak fluxes higher than 10^{16} n/cm^2/s and proposals as for the European Spallation Source (ESS) show that even several 10^{17} n/cm^2/s should be possible using a compressed beam of H$^-$ hitting a liquid Mercury target.

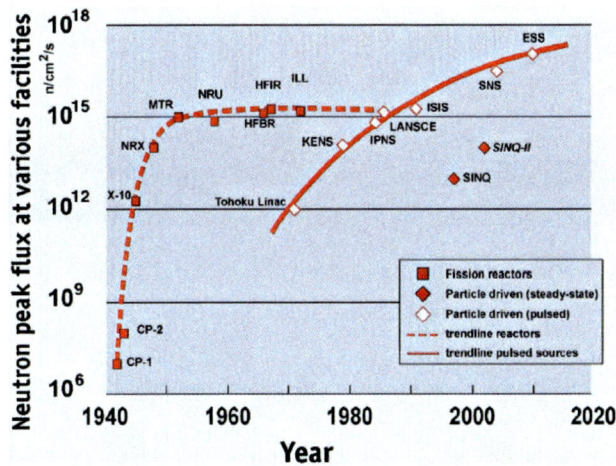

Fig 1: Development of innovative Neutron Sources as High Flux Reactors and Particle Driven Sources [3]

Materials Research for Fusion

The main scientific driver for materials research for fusion is the demand for plasma facing components and blankets for Iter (the international thermonuclear experimental reactor), where its is expected that the average neutron flux will be higher than 0.5 MW/m^2 and the average neutron fluence will be higher than 0.3 MWa/m^2[4]. The treatment systems for radioactive wastes generated by Iter and future fusion power plants shall be designed to minimize dispersion of radioactive materials. Various aspects of a future fusion power plant have been presented at a recent Technical Meeting on "First Generation of Fusion Power Plants: Design and Technology"[5].

One major future experiment on materials research for fusion components will be briefly introduced. The concept of the International Fusion Materials Irradiation facility (IFMIF) is based on the need of an intense neutron flux with an energy spectrum similar to the fusion neutron spectra. A 40 MeV proton beam hitting a liquid Lithium target could fulfill this request. The resulting neutron flux will be in the order of 10^{17} n/(100cm^{2}s) if two beams will hit the target simultaneously.[6] The neutron flux in the test area will provide a damage production of about 50 dpa/year in 0.1 L or about 20 dpa/year in 0.5 L (equivalent to 2 MW/m^2). IFMIF itself presents some significant materials challenges with its power and flux.

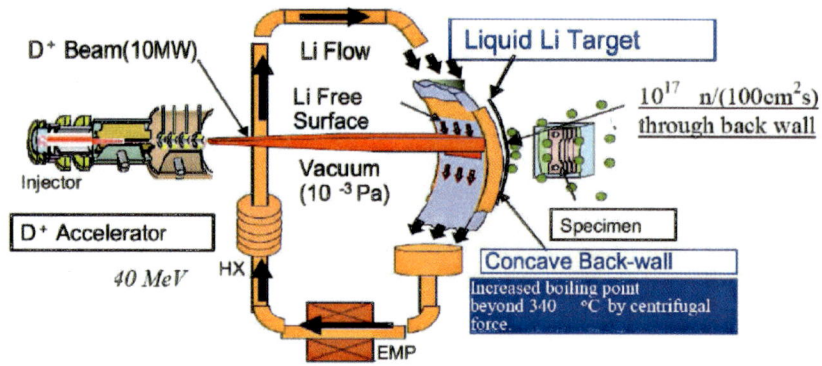

Figure 8. The IFMIF design concept

Fig. 2: The International Fusion Materials Irradiation Facility design concept

Further to the needs of the magnetic fusion experiments there are similar needs for inertial fusion experiments, using lasers or particle beams to drive the fusion process. One realization towards an inertial fusion power plant is FIREX in Japan, the "Fast ignition realization experiment" It is a very high temperature experiment driven by laser.

Following Iter, DEMO will be a power generating fusion reactor, but with flux and power more than 10 times Iter. There is an essential role for research reactors in support of the next generation of fission and fusion reactors; if member states identify a common research theme there is strong possibilities for an IAEA programme that will have broad impact in the industry. Figure 3 gives a recent summary for the present magnetic fusion devices. The injected power is at maximum about 10 MW for long pulse discharges, resulting in a total confined energy of about 1 GJ. Iter will operate at several 10 MW injected power, resulting in several 10 GJ. The situation for DEMO will be even more demanding. These issues on operation and high heat fluxes on plasma facing components are discussed at Technical Meetings on Steady State Operation [7].

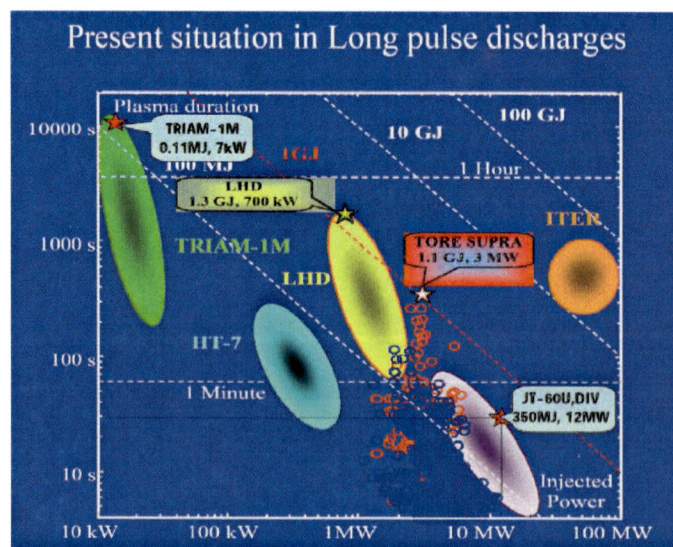

Fig. 3: Resulting energy in long pulse discharges in magnetic fusion devices as a parameter of injected power and pulse duration.[7]

4. SUMMARY

Material research at research reactors will be most important for future projects on Gen IV fission reactors and fusion power plants. The increasing demand on the neutron flux will lead to the use of spallation sources for specific materials research. Demands regarding the energy distribution of neutrons can be only fulfilled including specific moderators for special beamlines, or even specially build devices as IFMIF. New technologies will be needed to build the new neutron sources, but the research at these sources will result in advanced design of new nuclear power plants.

REFERENCES

[1]   Proceedings of the United Nations International Conference on the peaceful uses of atomic energy, Geneva (1955).

[2]   INTERNATIONAL ATOMIC ENERGY AGENCY, Development Opportunities for Small and Medium Scale Accelerator Driven Neutron Sources, IAEA-TECDOC-1439, IAEA, Vienna (2005).

[3]   RICHTER, D. (ed.), The ESS Project, Vol. II, ISBN 3-89336-302-5, (2002), p. 1-14.

[4]   INTERNATIONAL ATOMIC ENERGY AGENCY, ITER EDA Documentation Series No. 24, IAEA, Vienna (2002).

[5]   INTERNATIONAL ATOMIC ENERGY AGENCY, Proceedings Series, First Generation of Fusion Power Plants: Design and Technology, IAEA, Vienna (2006).

[6]   IFMIF International Team, IFMIF Comprehensive Design Report, IEA (2005).

[7]   Special issue on steady state operation (4TH IAEA Technical Meeting, Ahmedabad, India, 1–5 FEBRUARY 2005), Nucl. Fusion Vol **46**, no. 3 (2006).

LIST OF PARTICIPANTS

Boutard, J.L.	EFDA Close Supprt Unit-Garching, Boltzmanngasse 2, D-85748 Garching, Germany
Chandrasekharan, K.N.	Nuclear Fuels Group, Bhabha Atomic Research Centre, 400085 Mumbai, India
Harrison, R.	Institute for Materials & Engineering Science, ANSTO, Private Mailbag 1, Menai, NSW 2234, Australia
Inozemtsev, V.	International Atomic Energy Agency, Wagramer Strasse 5, A-1400, Vienna, Austria
Liu, T.	China Institute of Atomic Energy, Xinzhen St., PO Box 275 (33), Fangshan, Beijing 102413, China
Mank, G.	International Atomic Energy Agency, Wagramer Strasse 5, A-1400, Vienna, Austria
McIvor, A.	National Research Council of Canada, Canadian Neutron Beam Centre NRU Research Reactor, Chalk River, Ontario, Canada
Ohms, C.	Institutte for Energy, Joint Research Centre, European Commission, Westerduinweg 3, NL-1755 Petten, Netherlands
Paranjpe, S.K.	International Atomic Energy Agency, Wagramer Strasse 5, A-1400, Vienna, Austria
Paunoiu, C.	Institute of Nuclear Research Mioveni Campului Street 1 Casuta Postala 78 0300 Pitesti, Arges Romania
Schlamp, M.	Nuclear Engineering Dept., Centro Atomico Bariloche, Av. Bustillo 9500, 8400 Bariloche, Rio Negro, Argentina

Soares, A   International Atomic Energy Agency,
Wagramer Strasse 5
A-1400, Vienna, Austria

Van Der Schaaf, B.   NRG,
Petten, Netherlands

Wang, J.   Institute of Nuclear Energy Technology,
Tsinghua University,
100084 Beijing, China

IAEA

International Atomic Energy Agency

Where to order IAEA publications

In the following countries IAEA publications may be purchased from the sources listed below, or from major local booksellers. Payment may be made in local currency or with UNESCO coupons.

Australia
DA Information Services, 648 Whitehorse Road, Mitcham Victoria 3132
Telephone: +61 3 9210 7777 • Fax: +61 3 9210 7788
Email: service@dadirect.com.au • Web site: http://www.dadirect.com.au

Belgium
Jean de Lannoy, avenue du Roi 202, B-1190 Brussels
Telephone: +32 2 538 43 08 • Fax: +32 2 538 08 41
Email: jean.de.lannoy@infoboard.be • Web site: http://www.jean-de-lannoy.be

Canada
Bernan Associates, 4611-F Assembly Drive, Lanham, MD 20706-4391, USA
Telephone: 1-800-865-3457 • Fax: 1-800-865-3450
Email: order@bernan.com • Web site: http://www.bernan.com

Renouf Publishing Company Ltd., 1-5369 Canotek Rd., Ottawa, Ontario, K1J 9J3
Telephone: +613 745 2665 • Fax: +613 745 7660
Email: order.dept@renoufbooks.com • Web site: http://www.renoufbooks.com

China
IAEA Publications in Chinese: China Nuclear Energy Industry Corporation, Translation Section, P.O. Box 2103, Beijing

Czech Republic
Suweco CZ, S.R.O. Klecakova 347, 180 21 Praha 9
Telephone: +420 26603 5364 • Fax: +420 28482 1646
Email: nakup@suweco.cz • Web site: http://www.suweco.cz

Finland
Akateeminen Kirjakauppa, PL 128 (Keskuskatu 1), FIN-00101 Helsinki
Telephone: +358 9 121 41 • Fax: +358 9 121 4450
Email: akatilaus@akateeminen.com • Web site: http://www.akateeminen.com

France
Form-Edit, 5, rue Janssen, P.O. Box 25, F-75921 Paris Cedex 19
Telephone: +33 1 42 01 49 49 • Fax: +33 1 42 01 90 90 • Email: formedit@formedit.fr

Lavoisier SAS, 145 rue de Provigny, 94236 Cachan Cedex
Telephone: + 33 1 47 40 67 02 • Fax +33 1 47 40 67 02
Email: romuald.verrier@lavoisier.fr • Web site: http://www.lavoisier.fr

Germany
UNO-Verlag, Vertriebs- und Verlags GmbH, August-Bebel-Allee 6, D-53175 Bonn
Telephone: +49 02 28 949 02-0 • Fax: +49 02 28 949 02-22
Email: info@uno-verlag.de • Web site: http://www.uno-verlag.de

Hungary
Librotrade Ltd., Book Import, P.O. Box 126, H-1656 Budapest
Telephone: +36 1 257 7777 • Fax: +36 1 257 7472 • Email: books@librotrade.hu

India
Allied Publishers Group, 1st Floor, Dubash House, 15, J. N. Heredia Marg, Ballard Estate, Mumbai 400 001,
Telephone: +91 22 22617926/27 • Fax: +91 22 22617928
Email: alliedpl@vsnl.com • Web site: http://www.alliedpublishers.com

Bookwell, 2/72, Nirankari Colony, Delhi 110009
Telephone: +91 11 23268786, +91 11 23257264 • Fax: +91 11 23281315
Email: bookwell@vsnl.net

Italy
Libreria Scientifica Dott. Lucio di Biasio "AEIOU", Via Coronelli 6, I-20146 Milan
Telephone: +39 02 48 95 45 52 or 48 95 45 62 • Fax: +39 02 48 95 45 48

Japan
Maruzen Company, Ltd., 13-6 Nihonbashi, 3 chome, Chuo-ku, Tokyo 103-0027
Telephone: +81 3 3275 8582 • Fax: +81 3 3275 9072
Email: journal@maruzen.co.jp • Web site: http://www.maruzen.co.jp

Korea, Republic of
KINS Inc., Information Business Dept. Samho Bldg. 2nd Floor, 275-1 Yang Jae-dong SeoCho-G, Seoul 137-130
Telephone: +02 589 1740 • Fax: +02 589 1746
Email: sj8142@kins.co.kr • Web site: http://www.kins.co.kr

Netherlands
De Lindeboom Internationale Publicaties B.V., M.A. de Ruyterstraat 20A, NL-7482 BZ Haaksbergen
Telephone: +31 (0) 53 5740004 • Fax: +31 (0) 53 5729296
Email: books@delindeboom.com • Web site: http://www.delindeboom.com

Martinus Nijhoff International, Koraalrood 50, P.O. Box 1853, 2700 CZ Zoetermeer
Telephone: +31 793 684 400 • Fax: +31 793 615 698 • Email: info@nijhoff.nl • Web site: http://www.nijhoff.nl

Swets and Zeitlinger b.v., P.O. Box 830, 2160 SZ Lisse
Telephone: +31 252 435 111 • Fax: +31 252 415 888 • Email: infoho@swets.nl • Web site: http://www.swets.nl

New Zealand
DA Information Services, 648 Whitehorse Road, MITCHAM 3132, Australia
Telephone: +61 3 9210 7777 • Fax: +61 3 9210 7788
Email: service@dadirect.com.au • Web site: http://www.dadirect.com.au

Slovenia
Cankarjeva Zalozba d.d., Kopitarjeva 2, SI-1512 Ljubljana
Telephone: +386 1 432 31 44 • Fax: +386 1 230 14 35
Email: import.books@cankarjeva-z.si • Web site: http://www.cankarjeva-z.si/uvoz

Spain
Díaz de Santos, S.A., c/ Juan Bravo, 3A, E-28006 Madrid
Telephone: +34 91 781 94 80 • Fax: +34 91 575 55 63 • Email: compras@diazdesantos.es
carmela@diazdesantos.es • barcelona@diazdesantos.es • julio@diazdesantos.es
Web site: http://www.diazdesantos.es

United Kingdom
The Stationery Office Ltd, International Sales Agency, PO Box 29, Norwich, NR3 1 GN
Telephone (orders): +44 870 600 5552 • (enquiries): +44 207 873 8372 • Fax: +44 207 873 8203
Email (orders): book.orders@tso.co.uk • (enquiries): book.enquiries@tso.co.uk • Web site: http://www.tso.co.uk

On-line orders:
DELTA Int. Book Wholesalers Ltd., 39 Alexandra Road, Addlestone, Surrey, KT15 2PQ
Email: info@profbooks.com • Web site: http://www.profbooks.com

Books on the Environment:
Earthprint Ltd., P.O. Box 119, Stevenage SG1 4TP
Telephone: +44 1438748111 • Fax: +44 1438748844
Email: orders@earthprint.com • Web site: http://www.earthprint.com

United Nations (UN)
Dept. I004, Room DC2-0853, First Avenue at 46th Street, New York, N.Y. 10017, USA
Telephone: +800 253-9646 or +212 963-8302 • Fax: +212 963-3489
Email: publications@un.org • Web site: http://www.un.org

United States of America
Bernan Associates, 4611-F Assembly Drive, Lanham, MD 20706-4391
Telephone: 1-800-865-3457 • Fax: 1-800-865-3450
Email: order@bernan.com • Web site: http://www.bernan.com

Renouf Publishing Company Ltd., 812 Proctor Ave., Ogdensburg, NY, 13669
Telephone: +888 551 7470 (toll-free) • Fax: +888 568 8546 (toll-free)
Email: order.dept@renoufbooks.com • Web site: http://www.renoufbooks.com

Orders and requests for information may also be addressed directly to:

Sales and Promotion Unit, International Atomic Energy Agency
Wagramer Strasse 5, P.O. Box 100, A-1400 Vienna, Austria
Telephone: +43 1 2600 22529 (or 22530) • Fax: +43 1 2600 29302
Email: sales.publications@iaea.org • Web site: http://www.iaea.org/books